## About Island Press

Island Press is the only nonprofit organization in the United States whose principal purpose is the publication of books on environmental issues and natural resource management. We provide solutions-oriented information to professionals, public officials, business and community leaders, and concerned citizens who are shaping responses to environmental problems.

In 1999, Island Press celebrates its fifteenth anniversary as the leading provider of timely and practical books that take a multidisciplinary approach to critical environmental concerns. Our growing list of titles reflects our commitment to bringing the best of an expanding body of literature to the environmental community throughout North America and the world.

Support for Island Press is provided by The Jenifer Altman Foundation, The Bullitt Foundation, The Mary Flagler Cary Charitable Trust, The Nathan Cummings Foundation, The Geraldine R. Dodge Foundation, The Charles Engelhard Foundation, The Ford Foundation, The Vira I. Heinz Endowment, The W. Alton Jones Foundation, The John D. and Catherine T. MacArthur Foundation, The Andrew W. Mellon Foundation, The Charles Stewart Mott Foundation, The Curtis and Edith Munson Foundation, The National Fish and Wildlife Foundation, The National Science Foundation, The New-Land Foundation, The David and Lucile Packard Foundation, The Pew Charitable Trusts, The Surdna Foundation, The Winslow Foundation, and individual donors.

# Continental Conservation

# Continental Conservation
## Scientific Foundations of Regional Reserve Networks

*Edited by*
*Michael E. Soulé and John Terborgh*

THE WILDLANDS PROJECT

**ISLAND PRESS**
Washington, D.C. • Covelo, California

Library of Congress Cataloging-in-Publication Data
Continental conservation : scientific foundations of regional reserve
 networks / edited by Michael E. Soulé and John Terborgh.
      p.    cm.
   Includes bibliographical references.
   ISBN 1–55963–697–1 (cloth). — ISBN 1–55963–698–X (paper)
   1. Nature conservation.   2. Biological diversity conservation.
 I. Soulé, Michael E.   II. Terborgh, John, 1936–
 QH75.C685   1999                                        99–18457
 333.95'16—dc21                                          CIP

Printed on recycled, acid-free paper

Manufactured in the United States of America
10 9 8 7 6 5 4 3 2 1

# Contents

# *Preface*

The pages of a journal like *Ecology* . . . still packed with papers describing more and more sophisticated analyses applied to more and more trivial problems.

—Paul Ehrlich (1997)

The Wildlands Project (TWP) convened a workshop of thirty invited experts at the Rex Ranch near Tucson, Arizona, in November 1997 to discuss the science underlying the design and management of regional-scale networks of protected areas throughout North America. We designed the workshop with this book in mind. Not only were sessions scheduled with each of the main chapter titles as their theme, but participants were encouraged to shift from session to session as they saw fit. The session leaders assigned writing tasks to participants, and most chapters were outlined and begun before the end of the workshop. The rest of the writing was orchestrated by the chapter editors (senior authors) over the following nine months or so.

The authors are affiliated with many institutions, agencies, and conservation organizations and were selected because of their expertise and conservation interests. They do not necessarily endorse TWP. Among those present who are affiliated with TWP, either as staff, officers, or board members, were Steve Gatewood, David Johns, and Brian J. Miller. We ourselves, Michael E. Soulé and John Terborgh, are on TWP's board of directors and coordinated the workshop.

The Wildlands Project advocates a particular way of doing conservation: it stresses the design, implementation, and management of effective, regional networks of protected areas. One aspect of the workshop—apparent in the following chapters—was a critique of TWP's approach. Claims of objectivity are always suspect, but we have done our best to deal fairly with controversial issues. Moreover, it would be self-defeating for us to promote theories that misrepresent the best science available or to pretend that a certain approach to protecting nature is the last word. Our goal, however, was not merely to produce a handbook for advocates of TWP's approach. We believe that the conservation community as a whole will benefit from the science presented here. Our intent is to further the conservation of biodiversity—not to promote a particular style or agenda for accomplishing this.

Why would thirty busy scientists spend several days at a workshop devoted to synthesizing the science of regional and continental nature protection? Perhaps they came because they care about wildness and nature. Virtually every naturalist, ecologist, and field biologist in the world is saddened by the transformation and degradation of so much of the earth's natural areas. But a reasoned reluctance inhibits many scientists from becoming advocates, even for something as valuable as nature. In normal times, this kind of austere objectivity is useful to the advancement of knowledge. Even now, the majority of academic scientists are appalled by the thought of scientific activism under almost any circumstances. But when the very survival of something precious is at stake, aloofness may serve a lesser good than engagement. Such a time came for nuclear physicists during the Second World War. Paul Ehrlich believes that the present is such a time for biologists.

Why? Civilization has set aside only 5 or 6 percent of the earth's land in relatively strict protected areas for nature—for the 30 million or so species that depend on undeveloped lands and waters. This is not enough. Sadly, though, few of us rarely question our right to dominate nature—even to destroy it—particularly if it benefits the human economy. When David Ehrenfeld titled one of his books *The Arrogance of Humanism,* he meant to encourage conservationists (and, by association, conservation biologists) to educate society about the consequences of such actions, and about hubris.

But just as religion has not eradicated sin, conservation will not succeed in ending this extinction crisis simply by preaching about the destructive momentum of civilization. Yet conservation is making a difference—probably because so many people love nature. Much has been protected already by part-time activists, professional conservationists, hunters and anglers, agency personnel, a few enlightened politicians, generations of nature lovers, hundreds of committed scientists, thousands of courageous protesters, and multitudes of reluctant letter writers. But this is not enough. Humanity is alarmingly fecund: there are too many mouths to feed and too many dreams unrealized. And the ideological enemies of conservation will not rest until the last vestiges of creation are "improved."

Tired of being on the losing side of this sad game of endless appeals, compromise, and attrition, a group of conservation scientists and activists met in November 1991 to design a more effective way to protect nature, wilderness, and biodiversity. The principal rule of this approach is to honestly and boldly say what is necessary to save living nature in North America—how much land is required, where it is, and how the implementation should be phased over the coming decades.

The immediate objective is to produce—and then implement—map-based proposals for an effective network of nature reserves throughout North America. The elements of the networks include large, wilderness-like core areas linked by landscape corridors to facilitate natural flows. The network should be buffered, where appropriate, by lands that may also serve economic objectives. The science behind these projects is the subject of this volume.

A lot of people and organizations have helped to produce this book. Those who provided funding for the workshop and the editorial work include the Compton Foundation, the George Gund Foundation, The Money/Arenz Foundation, and an anonymous donor. Special thanks to the chapter editors for their spirit of cooperation and hard work. For their generous contributions of time, we also thank Dave Foreman, James A. Estes, Martha Groom, Richard L. Knight, Brian J. Miller, Katherine Ralls, Anatola Ondricek, and L. Scott Mills. Many people have helped in other ways, as well, including Hillary Oppman, Jennifer Dastrup, Barbara Dugelby, and Andy Robinson of TWP.

MICHAEL E. SOULÉ
JOHN TERBORGH

# 1 The Policy and Science of Regional Conservation

*Michael E. Soulé and John Terborgh*

This book addresses conservation at the continental scale by providing conservationists and biologists with the latest scientific principles for protecting living nature in whole regions and continents. But why is there suddenly a need for a new synthesis of conservation biology applicable at these large scales?

## Thinking Bigger

Nature conservation has largely been an ad hoc process. Tracts of land have been protected from development for their scenic value, or because they contain impressive concentrations of wildlife, or less often because they harbor rare species or notable biological diversity. This ad hoc approach to biological conservation has left Canada, the United States, Mexico, and most other countries with highly fragmented systems of parks and reserves in which some elements of the native biota are over-represented and others are not represented at all. Not only are most of the protected areas too small, but they are also isolated from other protected areas by agricultural lands, freeways, industrial zones, or other unnatural environments that are inimical to the large majority of native species. The resulting fragmentation of the natural environment severely threatens biological diversity. Modern conservation biology provides abundant evidence that small preserves, over the years, lose species. Even the largest national parks in western North America are too small to maintain all their larger mammals (Newmark 1995).

Many species cannot long survive in small habitat remnants because the ecological processes that stabilize their populations have been distorted or disrupted. Small areas are incapable of supporting the full spectrum of processes that sustain diversity—most notably the population stabilization of prey provided by top carnivores. Where crucial processes such as predation are distorted or absent, ecological systems often collapse with the loss of large numbers of species. The lesson learned from experience with small preserves is this: To be effective, biological conservation must be planned and implemented on large spatial scales. Conservation biologists have learned that nature and wildness cannot be saved by protecting a piece here and a piece there.

This book expounds the scientific basis for applying conservation principles at spatial scales encompassing whole regions and continents. A new conservation paradigm is needed because the protection of wilderness and biodiversity, including viable populations of large carnivores, is now recognized as a more formidable challenge than had been anticipated only a decade or so ago. Not only was scientific theory for protecting biodiversity insufficiently developed, but many of the social theories that guided the policies of conservation organizations have proved to be flawed. (See Kramer et al. 1997; Terborgh 1999.)

We do not want to give the impression, however, that the approaches and prescriptions recommended in this volume are meant to replace current conservation practices. First, not every conservation goal can be—or should be—addressed at the regional or continental scale. The challenges at the local level are profound and important; they must not be neglected. It would be disastrous if everyone abandoned the conservation needs of their communities to work exclusively at the level emphasized in this book. And second, depending on the social or biological context, some regional approaches can be inappropriate. Where large predators never existed (islands), or where they are long extinct (Australia), or where the density and intensity of human occupation is so great that the reintroduction of large animals is precluded (much of Europe, parts of Asia), there may be less biological justification for regional and interregional connectivity. Moreover, the kinds and scale—temporal and spatial—of conservation projects envisioned in the following chapters require a certain level of political stability (Soulé 1991), a certain level of national wealth, a certain level of organization, and most certainly a degree of human dignity, human rights, and social justice (Ehrlich et al. 1995). Nonetheless, we believe that most of the principles elaborated in this book, mutatis mutandis, apply worldwide.

While granting that there is still a major role for conservation of smaller, more isolated wildlands at the local scale, it has been apparent for a quarter-century that connectivity and bigness of protected areas are ecological necessities (Frankel and Soulé 1981). Until recently, however, the scientific theory for the protection of biodiversity was insufficiently developed and synthesized to guide conservationists at this bigger scale of enterprise. This book aims to conceptualize the scientific basis for this campaign—networks of protected areas that extend through entire regions and weave together wildlands on entire continents. This chapter sets the stage by introducing the major themes and background of policy and science that motivate these prescriptions.

A sense of urgency exists among conservationists everywhere. Not only is the present situation grim, but the rates of habitat loss and fragmentation are certain to increase (Soulé and Sanjayan 1998), particularly in the tropics. The challenge is daunting. Yet the present contains many possibilities. Many wildlands can still be protected, especially in the less populated and wealthier nations. Nearly everywhere there are opportunities to restore degraded and derelict lands to conditions that support biological diversity and provide ecological services.

Success in this endeavor depends on two conditions: good science and the popularization of a compelling, practical, and inspiring vision. The book outlines the current body of science—the description, theory, and prescription for large-scale conservation projects as presented, for example, in Noss and Cooperrider (1994)—and describes one such motivating vision (*Wild Earth*, Special Issue 1, 1992; Foreman et al. 1992) in the Preface. Here let us simply observe that conservationists who are courageous enough to confront the inexorability of current trends have only one constructive recourse: to pause, put aside at times their defensive tactics, their pleading, appealing, dealing, and suing, and help fashion a radically honest, scientifically rigorous, alternative land-use agenda. A major barrier at present is the dominance of policies, worldwide, that dismiss the utility of protected areas and promote the notion of sustainable development.

A large body of research in conservation biology has shown that maintaining ecological structure, diversity, and resilience demands strict, large-scale protection of entire ecosystems (see Noss and Cooperrider 1994). There is evidence that roughly 50 percent of the land in a region needs to be protected in systems of linked core areas if the goal of preventing further anthropogenic extinctions is to be achieved (Soulé and Sanjayan 1998). But most attempts to enlarge nature's estate by increasing the amount of effectively protected lands have failed, both in the rich and the poor countries. The percentage of land protected from exploitation worldwide has increased only marginally—from 4 percent to just over 5 percent—during the last two decades. One reason for this has been a lack of consensus on the need for more and larger protected areas associated with a policy shift away from protection and toward development.

The zenith of land protection occurred during the 1960s and 1970s. A major landmark in the United States was the passage of the Wilderness Act in 1964. It succeeded in protecting some federal Forest Service lands in the West from intensive exploitation or development. But lobbying by natural resource industries and developers trimmed the amount of land protected from the 24 million hectares proposed originally to about 3.6 million—an area about half the size of Maine. And recently the anticonservationist U.S. Congress has stifled spending for land protection under the Land and Water Conservation Act.

Meanwhile in Africa and Asia, many colonies were achieving independence. In those headier days it was fashionable for national leaders to create protected areas, even if only on paper. These declarations met with little resistance because human populations were half of what they are today and the opponents of protected areas were less influential and less organized. Today with renewable resources being harvested at unsustainable rates almost everywhere, the opportunity cost of creating additional preserves is viewed by many politicians as being too high. Peru, for example, has declared a moratorium on creating any new protected areas in the Amazon until the entire region has been explored for fossil hydrocarbons.

On the social side, it was fashionable in the 1960s and 1970s to support international conservation by hiring, training, and outfitting park guards. But beginning in the early 1980s, the policy pendulum in international assistance began to swing away from park protection and enforcement toward local economic development. The new bandwagon became Sustainable Development. Promotion of sustainable development was a well-intentioned but naive strategy to link economic growth to nature protection. Today it is evident that the effort to protect life on earth is failing—despite all the outpouring of feel-good news releases about sustainable development, integrated conservation and development, community-based conservation, ecosystem management, and sustainable forest management. These harmony-based notions are inspiring indeed. Their benefits for nature, however, are more likely to bear fruit during the twenty-second century—after the waves of population and market globalization have passed—but too late to save most of wild nature.

# The Region in Conservation

What are the conditions necessary to save living nature and wildness, particularly in North and Central America? Are the current intensity and scale of conservation sufficient to forestall the prophecies of an anthropogenic mass extinction? Is the conservation movement sufficiently organized and funded? What are the ecological and genetic principles of sound conservation at realistic geographic scales? These questions constitute much of the contemporary conservation challenge. Several organizations are attempting to address these issues, including the integration of scientific principles into the practice of conservation. Here are the principal elements in a strategic analysis:

- The contemporary mass extinction is being caused by anthropogenic forces: habitat loss and fragmentation, overexploitation of large or economically valuable animals (terrestrial predators, primates, parrots, marine fishes) and sensitive ecosystems (aquatic ecosystems, old-growth forests, coral reefs), introduction of invasive alien species, pollution, and climate change. The pressures on wildlands caused by these forces are still increasing.
- The design of functional protected area networks requires knowledge of how ecosystems are regulated—particularly the role of large carnivores.
- The spatial distribution of nature reserves and wilderness should be based on the principles of biogeography, wildlife biology, ecosystem ecology, and macroecology, with attention given to the anthropogenic forces listed above.
- Measures must be put in place immediately to maintain critical species, ecosystems, and landscape connectivity. Otherwise, human population growth and conversion of landscapes to agriculture and other economic uses will render these solutions politically impossible.
- Saving biodiversity and wilderness will require conceptualization and implementation of projects on larger spatial and temporal scales than heretofore envisioned.
- Other solutions to the impending biodiversity crisis—those favored by the international organizations, including the agencies of the United Nations and lending institutions—are exacerbating the problem by continuing to propagate the myths of harmony with nature, sustainability, and the notion that economic development is a necessary precursor of conservation (Soulé 1995).

In response to this analysis, conservationists are called upon to reexamine the scientific principles and management guidelines that have been in vogue during the last two decades and begin to reshape them for use at larger spatial and temporal scales. The goal of this chapter is to set the stage by addressing a number of critical issues. While the focus is on North America, the chapters refer to other continents; many of the principles are universal.

## A Fresh Approach

In its mission statement The Wildlands Project (TWP) diagnosed the current conservation challenge in spatial and temporal terms and emphasized the need to think and plan on grander geographic scales—regional networks of conservation lands—and to implement these networks over long periods of time if necessary (Foreman et al. 1992). While noting that today's parks, wilderness areas, and wildlife refuges preserve scenery and provide recreational opportunities, the mission statement recognizes that "they are too small, too isolated, and represent too few types of ecosystems to perpetuate the biodiversity of the continent." Underlining the implied dimensions of this diagnosis, the statement calls for "vast landscapes without roads, dams, motorized vehicles, . . . where evolutionary and ecological processes . . . can continue."

The infrastructure of this conservation strategy would comprise systems of core reserves, insulated in places by buffers, and linked within and between regions by ecological corridors that allow natural movements—including dispersal of wide-ranging species for genetic exchange between populations and for the migration of animals in response to climate change. A forerunner of this model is the biosphere reserve of UNESCO's Man and Biosphere Program; see Noss (1992) for references. How important is this typology of cores, corridors, and buffer zones? Perhaps the most sublime moments in intellectual life are when old frameworks dissolve and new ones take form. This appears to be happening to this trilogy. Though it is still convenient to use these land-use constructs (and we retain them as chapter headings), they are becoming obsolete in many contexts.

Why is this? Above all, the definitions are becoming blurred. The reader will discover that core areas, particularly large ones, provide connectivity for many species and processes. Moreover, regional and interregional corridors provide core habitat for many species, even for top carnivores. And buffers, assuming they are well designed, can be restored or managed so that they function as corridors and core habitat for some species. Finally, buffers may become core-like, and parts of cores may even have to become buffer-like, if only for reasons of economic and political expediency.

Regardless of one's view of categorical rigidity, however, a premise of this overall conservation strategy is that land-use planning must happen at the regional level, must be participatory, and must occur on spatial and temporal scales never attempted before (Soulé 1992). In fact, the vision is explicitly continental: the goal is to join regional networks together to recreate the potential of movements and flows that were severed decades to centuries ago by land uses and policies favoring helter-skelter development and shortsighted exploitation of natural resources.

## The Need for a Regional Strategy

Several organizations, including The Nature Conservancy and the World Wildlife Fund in the United States and Canada, now support the planning of nature reserves on a scale encompassing entire regions. What do we mean by "region," and why has it become a major focus in conservation planning and implementation? According to the Oxford English Dictionary, a region is a realm (region derives from "regere" in the sense of "to rule") or country distinguished by certain natural features, by its climate, and by its flora and fauna. Classical examples of regions encompass a range of spatial dimension—from hilly places to entire continents. Indeed, Sclater noted in 1858 that "South America is the most peculiar of all the primary regions of the globe as to its ornithology."

Today the term "region" refers to some intermediate scale between locality and continent and is often defined by obvious topographic or biogeographic features such as vegetation. Thus we refer to the boreal forest region of Canada or to the Sonoran Desert region encompassing much of northwestern Mexico and part of the southwestern United States. A region is large: bigger than a landscape, bigger than any single national park, bigger than a mountain or lake (hence the Southern Rocky Mountains or the Great Lakes Region), and, therefore, bigger than the jurisdiction of most land management units, including many county, state, provincial, or even national governments. Planning for regions, then, requires the protection of biodiversity and wilderness on a scale that is broader than the dimensions of these traditional administrative elements (Clark and Minta 1994). The region as a geographic unit for conservation represents a ratcheting up of conservation planning.

Therefore a regional, let alone a continental, approach requires forms of design, implementation, and management that are qualitatively different from local conservation (such as single nature reserves or even national parks). A regional scope presents new possibilities. Projects that are impractical locally (such as the reintroduction of large carnivores) become feasible at a regional scale. Stand-replacing fires that may be catastrophic at the local or small-wildlands scale can be left to burn themselves out in a region. Problems of transboundary conservation—the brown (grizzly) bear in Canada and the United States, for example, and the jaguar and Mexican wolf in Mexico and the United States—appear formidable from the perspective of states but become tractable within a regional context. Conversely, projects that are feasible at the local scale (such as the complete elimination of exotic species by volunteer workers—say from a wetland or sand dune) are difficult if not impossible at a regional scale and alternatives must be considered. In summary, then, a regional approach must deal with the complexities of scale (Chapter 2) as they apply to process, species (Chapter 3), and restoration (Chapter 4).

## Continental Conservation

Do analogous problems exist at the continental scale in the context of nature conservation? Several conservation issues must be addressed at a bigger scale than the region. North America's peregrine falcons (*Falco peregrinus*), for example, were nearly extirpated by the egregious use of pesticides. Conservation biologists and falcon aficionados responded by mounting a successful continental effort to save them. The notion of continental scale is also meaningful in a biogeographical context. Entire families and genera of plants and animals are often restricted to individual continents, and conservation coordination may have to be continent-wide: the people concerned with such taxonomic groups must address continent-specific ecological and cultural factors. The macaws, for instance, a group of large, New World parrots, are found only from Mexico to South America and are seriously threatened because of the pet trade and habitat destruction. In this case the remedies—conservation and political—must be administered on a continental scale.

The continent is also the geographic unit of relevance for many large-bodied, wide-ranging species. Moreover, a genetically and demographically viable population of these animals may number in the thousands of individuals—thus requiring areas comprising many regions to provide sufficient habitat. For such species, the continent (or a large part of one) is the appropriate conservation scale. Thinly spread species of large carnivores like the brown bear, the puma (*Felis concolor*), the jaguar (*Felis onca*), and the wolverine (*Gulo gulo*) are often not viable, genetically or demographically, within single regions, even those as large as the Sierra Madres, the Adirondacks, or the northern Cascades. Viable populations of such animals require vast areas spanning many regions. And because such creatures are often keystone players (Mills et al. 1992; Power et al. 1996; Chapter 3) in ecosystems, their conservation is a supraregional or continental issue.

## The Scale of Things

So far we have made the assumption that you—the reader—are comfortable "zooming" from, say, your backyard to your entire continent, and that you have some understanding of the different problems faced by individual creatures and entire species that are relevant at these different geographic scales. Indeed, issues of scale are ubiquitous in conservation, particularly as we consider regional networks and continental conservation (Chapter 2). A major emphasis in this book is the need to think and plan on scales that transcend traditional political boundaries (counties, states, provinces, nations) and familiar spans of time (lifetimes).

Most planners and politicians are hemmed in by narrow jurisdictional, bureaucratic, and political horizons, including terms of office. Such strictures are too limiting for conservation. The conservationist must cultivate the capacity to shift smoothly from, say, the needs of individual plants and animals—on seasonal and annual scales—to the temporal and spatial criteria for long-term population viability (centuries and millennia) and the long-distance interchange of material and energy between marine and terrestrial ecosystems (Chapter 6). Managers, too, must develop a facility with scale. At the local scale the exotics problem might be manageable, for example, depending on the vigilance and commitment of local authorities and the interested public (Chapter 4). At the national level, however, the management of an alien species may involve complex trade negotiations with importers, issues of sovereignty, legislation, and public education.

Issues of scale can also create conflicts. The managers of a local nature reserve—say one established to protect a particular marsh, wetland, or patch of prairie—may not feel compelled to plan for the viability of wide-ranging or migratory species such as cranes, caribou, cougars, or coyotes. "This," they will say, "is somebody else's problem." Though such a cavalier dismissal of certain species should never be condoned, it is understandable. The problem is most evident where large carnivores are the issue. Fear of being mauled and eaten by these animals (the "Jaws" syndrome) is nearly universal—notwithstanding all the statistical evidence about the frequency of such attacks compared, say, to injury and death in the home, in the automobile, from lightning, or even from encounters with bees or deer.

In many cases conservation activists shun the issue of predators because they worry about opposition from ranchers and worried parents. Compliance with such negative anthropocentric attitudes, however, requires the rejection of the scientific arguments for the roles of top carnivores (Chapter 3), not to mention the dismissal of the prime conservation goal: the return of viable, healthy populations of all native species (Noss 1992). And "all" means all—even those animals that may not be convenient or beautiful. It is the responsibility of conservationists to educate themselves and their human communities about the ecological importance of predators, not duck the issue. Moreover, local and regional conservationists cannot ignore their responsibility to the whole system: each planning and implementation group must bear its share of the regional responsibility for all species and ecological processes.

## Top-Down Regulation

A central issue in this volume—a theme that emerges in every chapter— is the important role played by keystone species, particularly wide-ranging carnivores, in ensuring the viability of protected areas. In fact, research is showing that the rules guiding the architecture of regional conservation networks are, to a considerable degree, determined by the habitat requirements of keystone species (Chapters 3 and 5).

The scale and dimensions of a conservation network are often dictated by the needs of wide-ranging species, particularly top predators. Although a key role for top predators had been dismissed as recently as 1996 (Polis and Strong 1996), there is increasing evidence that the absence of large carnivores can initiate cascading perturbations through the trophic web—disturbances that often manifest as superabundant populations of herbivores and medium-sized predators. These perturbations, in turn, may cause reproductive failure and local extinction of plant species and prey, including birds, reptiles, amphibians, and rodents. A complementary hypothesis holds that productivity-limited ecosystems—those regulated by the growth and biomass of plants—are not common in the presence of the complete set of native carnivores, except, perhaps, in arid regions.

Because the viability of ecosystems may often depend on the viability of species whose interactions regulate the systems, it follows that the size of a system and its actual, on-the-ground, configuration of boundaries and corridors must accommodate the needs of a critical handful of highly interactive species. Because these species often include the large carnivores—which are both admired and feared by human beings but embody the essence of wilderness—the goal of maintaining viable populations of keystone species, particularly large carnivores, has been referred to as "rewilding" (Soulé and Noss 1998).

Rewilding is the latest element in the history of scientific conservation. It does not, however, stand alone. Rather, it complements other approaches for designing regional networks of nature protection by contributing an independent justification for bigness and connectivity. And like certain other methodologies, it facilitates design and management of protected areas because it obviates the need to consider every species in detail (Chapter 5). Thus rewilding is both an end (because of our duty to repair past mistakes in management) and a means by which the viability of conservation units is achieved. This unusual conjunction of means and ends is, perhaps, the most intellectually compelling feature of rewilding.

Prominent among the complementary methodologies for the design of reserve networks are the "coarse-filter" approach and the consideration of disturbance dynamics (Chapter 5). When the conservation mission includes the protection of all elements of biodiversity, for example, including native species, it is virtually imperative to incorporate representatives of each type of biotic community that occurs in the region. Still, the representation of all biotic elements is no guarantee that these elements will persist. In most cases, the determination of minimum viable areas for vegetation or habitat types must be based on the needs of particular species, usually the wide-ranging predators, and on the temporal/spatial scale of major classes of disturbance. Thus there is an important distinction to be made between protection (inclusion in a protected area) and preservation (long-term persistence of protected species). At present there is no scientific formula for determining minimum viable areas for the preservation of vegetation or habitat types, except by reference to the needs of particular species.

## Rewilding and Connectivity

Nature is now in pieces, and rewilding is a justification for restoring connectivity on a regional or landscape level. This is because the remaining remnants of habitat are nearly always too small and too isolated to provide sufficient habitat for the top carnivores. Therefore, except in undenatured "frontier" regions such as much of Canada north of the 50th parallel (54th parallel in British Columbia) and Alaska, it is necessary to plan and implement systems that will restore connectivity. Connectivity also maintains or restores ecological processes that have been severed or severely constricted by human activities (Chapter 6).

As inhabitants of a world in pieces, we need to remind ourselves that connectivity is not just another goal of conservation: it is the natural state of things. Connectivity, therefore, is a sine qua non for conservation. Originally nature was connected on the scale of continents, though this certainly does not justify the willy-nilly creation of artificial corridors. Rarely will haphazard reassembly of habitat patches restore an ecologically viable landscape. Connectivity, per se, is not the goal we advocate (Chapter 6). The goal is to reverse the terrible consequences of fragmentation at the habitat and landscape scale—to restore the effective exchange of individuals and materials among sites for genetic maintenance, for demographic stability, for migration, and for the sake of other ecological processes. But, as critics have pointed out, connectivity is not analogous to an elastic bandage: one size does not fit all. Nor does one kind of connectivity solve all problems. The kind and scale of connectivity must fit the context and address the goals of the project at hand.

The term "habitat fragmentation" evokes a static, two-dimensional notion—a kind of poorly constructed checkerboard of habitat islands on a background of denatured land converted to economic uses. The metaphor of fragmentation, therefore, fails to convey the quivering isolation of animals, energy, and nutrients in a degraded landscape. Wild nature is full of movement and interchange, often on a scale of hundreds and thousands of kilometers. It is this dynamic element of nature that the notion of fragmentation fails to capture. In effect, each isolated remnant of nature is caught in a tightening tourniquet of civilization. Initially the tourniquet is made of roads or small subsistence farms. Gradually the constraining band broadens outward as more habitat is converted to farms, clearcuts, and villages. Eventually highways, dams, intensive agriculture, and cities become impervious barriers in the landscape, permeable only to aerial flyers and drifters. Some of these flyers, like the long-nosed bats described in Chapter 6, perform critical services by moving between isolated ecosystems and knitting together island habitats.

Not all so-called corridors benefit nature. Some—such as utility corridors, equestrian and bicycle paths, and greenways—facilitate the travel of invasive exotics, human beings, and their pets, seriously hampering the viability and movements of native plants, animals, and the vital flows of materials and energy that constitute the circulation of wildlands

(Moran 1994). Nevertheless, restoration of ecological connectivity must be a ubiquitous conservation activity in the temperate and tropical regions.

### Rewilding and Restoration

Ecological restoration on a regional scale is a new endeavor (Chapter 4). Until recently, restoration ecology has emphasized the repair of local wounds such as those caused by mining, agriculture, grazing, and invasive, exotic species. Now we face a larger challenge—including the need to reverse decades of fire suppression in forests, nearly a century of predator extirpation, overgrazing, and other abuses to wildlands.

The field of ecological restoration has not normally been concerned with biodiversity or conservation per se. In fact, native species have been a secondary objective in most restoration projects. The focus in restoration has been the recovery of ecological *process,* particularly soil stabilization and productivity of polluted or derelict lands. But ecological processes are largely an epiphenomenon—they depend on the persistence of certain species (Shrader-Frechette and McCoy 1993). The restoration of native species diversity, including large carnivores, will require theory and experimentation on a larger scale. Coordination of projects is essential, so that the cumulative impacts of restoration projects promote the overall conservation goals of the regional network.

All this requires a significant shift in ecological perspective: a new paradigm in restoration ecology. Most restorationists, for example, do not think in terms of top-down regulation of ecosystems; rather, the dominant paradigm is bottom-up, driven by succession in plant communities. At the regional scale, however, the emphasis often may be on the recovery of large carnivores and their trophic base of ungulates.

## Maps: The Land-Use Agenda

Biological conservation, rewilding, and restoration are complementary tools for nature protection. Their implementation, however, is virtual rather than real until positive changes occur on the ground. Such change is almost always guided by maps.

Maps stimulate desires—for territory, for natural resources, for real estate development, even for conservation. Therefore, the ideology of those who produce land-use maps is important. If developers are the only people mapping the land's future use, then they control the land-use agenda. Indeed, the dominant ideology of land use is development. But the impact of development on nature is not "development" (less entropy or more order)—it is more entropy. For biodiversity, development is really de-development or denaturation (Soulé 1991).

One of the reasons why economic development usually trumps nature protection is that promoters of growth have a corner on the production of maps used by jurisdictions and agencies that make land-use decisions—decisions that define the future of the land. Development maps promise to deliver social or economic benefits. In the world of planning, land-use maps are promissory notes for economic growth—increased profits and more local employment. Conservationists and others who oppose unending growth and development are often accused of opposing progress and forced on the political defensive. This is not acceptable.

If maps are the agenda, then conservationists must enter the mapping arena. They must begin producing attractive alternatives—maps that also promise a social good: the benefits of wilderness and nature protection. Such maps must be honest, scientifically rigorous, and inspiring so that the hegemony of the growth myth can be offset by the notion that the long-term interests of society are well served by a sufficient system of protected areas. Science and maps, when allied, constitute a powerful partnership.

Good conservation maps require scientific knowledge as well as technologies like geographic information systems. This troubles some conservationists who believe that science and technology contaminate the aesthetic and spiritual arguments for nature and wilderness conservation (McCloskey 1996–1997). Other critics simply dislike science per se or are revolted by invasive research and management interventions—such as the use of radio collars (Turner 1996). Not all scientists (nor all conservationists) are saints, of course, but science and advocacy must become allies in the defense of nature. The essential role played by conservation science today is made explicit in the following chapters.

## A Vision and a Challenge

The new field of regional and continental conservation offers many opportunities for research, synthesis, and application. Time, however, is running out for nature as the world enters an era of rapid, chaotic, dissipative change—nature's end game. In the short run, the most noteworthy of these trends include the globalization of commerce and the statutory and technological facilitation of trade and resource extraction. Except in a few cold places, ecosystems will be over-whelmed as the human population and economy continue to grow and expand.

What attitudes and behavioral changes will be required? Every field biologist knows the world is in crisis. And crises call forth new roles, including that of the scientist/activist—the articulate, committed expert. Deftness will be required of those who have chosen conservation of nature as their mission. The most skillful will learn to abandon favorite theories, comfortable opinions, cherished classifications, and the fear of controversial policies—such as the reintroduction of large carnivores. And, of course, an inspiring vision is essential. In the frenetic, noisy years ahead, only such visions will attract attention and kindle hope. Finally, conservation on the ground must replace the repetitive cycle of confer-ences, reports, recommendations to governments, and ineffective treaties.

Continental conservation is already intellectually challenging. But new, more comprehensive methodologies and theoretical constructs will be demanded. There will, however, be less tolerance of models that fail to reflect both biological and social realities. The litmus test of ideas is no longer just elegance; it is pragmatism. For good or bad, the question must be: does it work, and for how long? Social scientists can help to answer these questions by informing conservationists about cultural dif-ferences and opportunities, and by providing guidance for implementa-tion, including how to address conflicts in values and such problems as cross-border (international) cooperation.

But biologists themselves are not necessarily inept when it comes to implementing conservation on the ground. Organizations such as the Wildlife Conservation Society and Conservation International, as well as the careers of many dedicated tropical biologists, have repeatedly demonstrated that the long-term presence of a scientist/activist can make a profound difference in a region. A perennial, personal commit-ment to a place appears to accomplish more than the infusion of large dollops of dollars.

The next three chapters explore the conceptual landscape of regional and continental conservation. Chapter 2 surveys the complexities of scale. Chapter 3 considers the significance of species interactions and ecological processes. Chapter 4 addresses the role of restoration at the regional scale. The scientific details of designing and implementing regional conservation in core areas (Chapter 5), in corridors (Chapter 6), and in buffer zones (Chapter 7) constitute the balance of this volume.

Chapter 8 urges the reader, whether scientist or advocate, to commit more time to the challenge of saving nature. Because it will soon be too late (Ehrlich 1997: 49, 91), no other generation will ever face this challenge with such urgency.

# References

Clark, T. W., and S. C. Minta. 1994. *Greater Yellowstone's future.* Moose, Wyo.: Homestead.

Ehrlich, P. R. 1997. *A world of wounds: Ecologists and the human dilemma.* Oldendorf/Luhe, Germany: Ecology Institute.

Ehrlich, P. R., A. H. Ehrlich, and G. C. Daily. 1995. *The stork and the plow: The equity answer to the human dilemma.* New York: Putnam.

Foreman, D., J. Davis, D. Johns, R. Noss, and M. E. Soulé. 1992. The Wildlands Project mission statement. *Wild Earth* (special issue): 2–3.

Frankel, O. H., and M. E. Soulé. 1981. *Conservation and evolution.* Cambridge: Cambridge University Press.

Kramer, R., C. van Schaik, and J. Johnson. 1997. *Last stand: Protected areas and the defense of tropical biodiversity.* Oxford: Oxford University Press.

McCloskey, M. 1996–1997. Conservation biologists challenge traditional nature protection organizations. *Wild Earth* 6(4):67–70.

Mills, L. S., M. E. Soulé, and D. Doak. 1992. The history and current status of the keystone species concept. *BioScience* 43:219–224.

Moran, K. K. 1994. Wildlife corridors and pipeline corridors: A comparative analysis. In K. G. Hay (ed.), *Greenways, wildlife, and natural gas pipeline corridors: New partnerships for multiple use.* Arlington, Va.: The Conservation Fund.

Newmark, W. D. 1995. Extinction of mammal populations in western North American national parks. *Conservation Biology* 9:512–526.

Noss, R. F. 1992. The Wildlands Project: Land conservation strategy. *Wild Earth* (special issue): 1:10–25.

Noss, R. F., and A. Cooperrider. 1994. *Saving nature's legacy: Protecting and restoring biodiversity.* Washington, D.C.: Defenders of Wildlife and Island Press.

Polis, G. A., and D. R. Strong. 1996. Food web complexity and community dynamics. *American Naturalist* 147:813–846.

Power, M. E., D. Tilman, J. A. Estes, B. A. Menge, W. J. Bond, L. S. Mills, G. Daily, J. C. Castilla, J. Lubchenco, and R. T. Paine. 1996. Challenges in the quest for keystones. *BioScience* 46:609–620.

Shrader-Frechette, K. S., and E. D. McCoy. 1993. *Method in ecology: Strategies for conservation.* Cambridge: Cambridge University Press.

Soulé, M. E. 1991. Conservation: Tactics for a constant crisis. *Science* 253: 744–750.

———. 1992. A vision for the meantime. *Wild Earth* (special issue): 7–8.

———. 1995. The social siege of nature. In M. E. Soulé and G. Lease (eds.), *Reinventing nature?: Responses to postmodern deconstruction.* Washington, D.C.: Island Press.

Soulé, M. E., and R. Noss. 1998. Rewilding and biodiversity as complementary tools for continental conservation. *Wild Earth* 8(3):18–28.

Soulé, M. E., and M. Sanjayan. 1998. Conservation targets: Do they help? *Science* 279:2060–2061.

Terborgh, J. 1999. *Requiem for nature.* Washington, D.C.: Island Press.

Turner, J. 1996. *The abstract wild.* Tucson: University of Arizona Press.

# 2

## The Issue of Scale in Selecting and Designing Biological Reserves

*J. Michael Scott, Elliott A. Norse, Héctor Arita, Andy Dobson, James A. Estes, Mercedes Foster, Barrie Gilbert, Deborah B. Jensen, Richard L. Knight, David Mattson, and Michael E. Soulé*

People aspire to conservation goals that range from the highly provincial to the universal. While local groups grapple with reclaiming New York City's "wilderness parks" from exotic weeds, statesmen negotiate global policies for limiting trade in endangered species and the protection of tropical forests. Somewhere toward the upper end of this spectrum of conservation issues are initiatives such as the Yellowstone-to-Yukon Project that aspire to protect, conserve, and restore biodiversity and wilderness throughout the northern Rocky Mountains. Encouraging such initiatives as Yellowstone-to-Yukon, The Wildlands Project seeks to "protect and restore the ecological richness and native biodiversity of North America." Such efforts have been launched because present-day conservation reserves have failed to prevent the loss of species (Newmark 1987, 1995). The goals of these large-area projects include the establishment of a continental network of protected areas large enough to support viable populations of even the largest vertebrates and to sustain landscape-level ecological processes that ensure the diversity and resilience of ecosystems. (See Chapters 1 and 5 for a thorough discussion of goals.)

## Thinking About Scale

To select and design reserves to achieve these ends requires a perspective that spans nations and centuries. These steps also require information such as satellite images that can be used to view and map wildlands and the surrounding matrix of farms, rangelands, exploited forests, and urban landscapes in which they are usually embedded. Such a macroscopic view of the world (Brown and Maurer 1989; Jennings and Scott 1993; Brown 1995) establishes the vital ecological and cultural contexts for additional studies of species, communities, and ecological processes that must be conducted at increasingly finer scales of spatial and temporal resolution. Thus a reserve designer must think about wildlands on many scales—from the most local, say a critical wetland, to the most regional, say the reintroduction of grizzly bears (*Ursus arctos*) into a 40,000-square-kilometer transnational core reserve. Clearly the design and management of regional networks of conservation reserves must occur at spatial and temporal scales that transcend the normal human life span and extend beyond traditional political boundaries.

In any ecological endeavor it is critical that the scale be appropriate to the question being asked (Kareiva and Anderson 1988; Wiens et al. 1986; Wiens 1989). To learn which conditions are needed to ensure the survival of jaguars (*Panthera onca*) along the United States and Mexican border, for example, one would focus on populations and population processes such as immigration, emigration, mortality rates, and recruitment—that is, on phenomena that operate over time periods and areas that transcend lifetimes and home ranges of individual cats. It would not suffice, for instance, to study only the movements of particular jaguars. But if one were interested in the effects of poaching deer on the physiology and reproduction of female jaguars, the focus would be on individual females—their physiological status.

To be successful, whether in action or inquiry, conservationists need well-tested operational guidelines that are appropriate for regional and continental scales and over periods of a hundred years or more. Without such a frame of reference, conservation efforts and scientific discourse typically founder because tactics that work well on a local, short-term scale are often inadequate at the regional scale. It would not be prudent, for example, to propose the reintroduction of wolves into an urban park. The scales of the park and the needs of wolves are incompatible—not to mention the political resistance to such an intervention. Similarly, volunteers may be effective in controlling an exotic species at the habitat patch and local levels, such as the removal by hand of French broom or pampas grass, but this approach seldom works with a regional scourge such as Russian knapweed or cheatgrass. In other words, we need to incorporate hierarchical thinking into our studies and planning (Allen and Starr 1982; O'Neill et al. 1986).

In this chapter we describe scale-related considerations in the identification, selection, and design of conservation reserve networks (Scott and Csuti 1997). Scale issues relevant to wildlife corridors are discussed in Chapter 6. We use the terms "broad scale" and "fine scale" to denote patterns and processes that are exhibited over large and small areas, respectively. We avoid using the terms "large scale" and "smallscale," whose meanings in the common vernacular differ from the jargon of geographers.

A major goal of broad-scale conservation projects is the regional or continental protection of living nature, or biodiversity. Biodiversity is the "variety and variability among living organisms and the ecological complexes in which they occur" (U.S. Office of Technology Assessment 1987). The term encompasses not only all species everywhere, but the variations in the composition, structure, and functional process of the ecosystems in which they live. Noss (1990) described an integrated biological hierarchy for biodiversity that recognized four components: genetic; population-species; community-ecosystem; and landscape or regional. When creating a reserve complex to preserve biodiversity, the various scales of biological and ecological organization must be fully considered. Each level of organization demands a different way, or scale, of thinking about nature.

*Individual requirements.* The needs of individuals are different from those of populations. They fall into four general categories: nutrients (energy, micronutrients, water); shelter (against energy loss); security (against predation or herbivory); and protection (from pathogens, parasites, poisons, and other chemical pollutants). The biological needs of many creatures also vary according to the stage in the life cycle. Most species of frogs, for example, are terrestrial as adults but require aquatic habitats for reproduction, larval development, and metamorphosis. Habitat and food resources for many insects also differ significantly between larvae and adults. Postmetamorphic adults of invertebrates and fishes in coastal marine systems may be highly sedentary whereas their larvae often move vast distances across the oceanic realm between fertilization and settlement (Roughgarden et al. 1988). Nomadic and migratory animals may range over huge areas in the course of a year although occupying only a limited portion of that area at any one time. These species will require corridors to link areas of seasonal use. (See Chapter 6.)

*Population-level requirements.* Protected areas must be large enough to support groups that can maintain genetic variability sufficient to prevent inbreeding depression and resist local population fluctuations. Multiple populations increase the likelihood there will always be a source of dispersing individuals to recolonize areas after local extinctions. Species and populations cannot survive in the absence of habitat, a place to live.

*Community-level requirements.* Animals and plants depend on other organisms. A community, therefore, needs to sustain viable populations of functionally associated species such as plants, pollinators, and seed dispersers, predators and prey, or parasites and hosts (Buchman and Nabhan 1996). Reserves must also be large enough to maintain nutrient cycles and ensure the distribution of seeds, spores, and other propagules. The presence of top predators may also be a community-level requirement.

*Landscapes or ecosystems.* Single reserves must be large enough to maintain populations of associated species to buffer against environmental perturbations such as flood, fire, and drought (or be part of a reserve system that accomplishes the same thing).

## The Use of Scale in Reserve Design

Anyone contemplating the protection of nature must familiarize himself with a variety of scale phenomena, including the scale of animal movements and the full range of spatial and temporal scales. Moreover, the conservationist must gain some facility in mentally moving between scales. In this section, we survey the needs of focal species, the relationship of scale to the sizes and geographic distribution of nature reserves, and how both short-term and long-term environmental variation can influence the selection and design of protected areas.

### Focal Species and Scale

The initial step in designing a reserve network is to identify its primary objectives. One of the most common goals is "protection of the contained biodiversity." But biodiversity is such an all-inclusive concept that designing a system to account for all the specific needs of all the organisms is unworkable. To get around this problem, conservationists generally focus their efforts on some subset of biodiversity: a surrogate for the ecosystem and all its species. Often the subset comprises a set of species of special concern or ecological importance, such as endemic, rare, keystone, indicator, umbrella, or flagship species. (See Frankel and Soulé 1981; Power et al. 1996; Meffe and Carroll 1997; Noss et al. 1995; Scott et al. 1993; Miller et al. 1999; Soulé and Noss 1998.)

The use of surrogates for biodiversity (such as species, species richness, or enduring physical feature of the environment) constitutes one approach to reserve identification and selection (Hunter et al.

1988). Another approach is the representation of all habitat or vegetation types (Shelford 1933; Caicco et al. 1995; Davis et al. 1995). This "coarse-filter" approach is discussed in Chapter 5. These approaches should not be seen as contradictory or competing, however (Soulé and Noss 1998). In all cases the intent is to garner benefits for a broad range of taxa while matching management to the analytical and monetary resources.

Whatever the organizing paradigm, reserves must be of a size and design appropriate to meet the needs of the target life forms or systems and their functionally associated species, ecological processes, and biological phenomena, at all life-history stages, at all seasons, and in the face of all reasonably likely environmental perturbations. In other words, the designer must be capable of working at all scales and levels at once. Often the designer will conclude that even a single large reserve is insufficient for the protection of biodiversity. In the case of large carnivores and wide-ranging migrants, for instance, a regional design process is typically required: a network of reserves containing the resources for different seasonal or life-stage requirements, not to mention long-term genetic fitness and evolutionary potential (Chapters 2, 5, and 6).

Animals that fly present special challenges. Whereas some birds hardly budge from their place of birth, the movements of others span the entire spatial scale from local to intercontinental. In a somewhat parallel fashion, some birds are reluctant to cross openings in the vegetation cover, such as rivers and roads. Others routinely forage over vast oceans or cross them on their annual breeding migrations (see Chapter 6). What may look like an ecological barrier to human beings may be permeable to many species (and vice versa). Our perceptual bias is one reason why biologists familiar with the behavior and needs of particular species must be involved at all stages of reserve design.

## Asymmetries and Hierarchies of Scale

Certain generalizations about reserve design are based on an inherent asymmetry of spatial scale—namely, the larger usually includes the smaller, a kind of one-way, or nested, hierarchy. This means, for example, that protection of areas that are large enough to maintain viable populations of wide-ranging species, such as wolves (*Canis lupus*) and wolverines (*Gulo gulo*), are frequently sufficient to protect many, if not most, species that require less space—the umbrella idea (Frankel and Soulé 1981; Meffe and Carroll 1997). The underlying premise is that vigorous populations of umbrella species translate into protection for the remaining biota. This assumption, however, has rarely been tested (Chapter 5;

Miller et al. 1999). The reverse does not work: a vigorous population of mice or pocket gophers does not guarantee the persistence of bears and mountain lions. Nevertheless even though small, individual reserves are not large enough to sustain a viable population of many large species, a connected network of small reserves may suffice (Beier 1993; Gilpin and Hanski 1991). Chapter 5 describes in detail the many approaches to reserve design.

Another asymmetry of scale is that a large reserve is likely to contain more species for a longer time than several small reserves that add up to the area of the large one. This generalization refers to the SLOSS (single large or several small) issue. Conservation biologists have debated whether a single large or several small reserves represents the best strategy for protecting biodiversity (Diamond 1975; Simberloff 1986; Simberloff and Abele 1976a, 1976b). Indeed, the inadequate consideration of spatial scale was at the heart of much of the SLOSS debate (Higgs and Usher 1980; Soulé and Simberloff 1986). Moreover, early arguments on both sides of this issue failed to recognize that the answer to the SLOSS question depends on what you are trying to protect, for how long, and where you are trying to protect it.

Another relevant scale consideration is the scope or size of the region. Suppose a small town decides to set aside an area of habitat along a creek as a local nature park. At this local scale, regardless of the creek selected, there is little difference in the species composition of similar size patches. The larger the park, of course, the larger the proportion of the regional local diversity of species it is likely to capture. But as the scale increases to the state and national levels, the similarity of species composition declines between patches of habitat of similar sizes—say between distant localities like Eugene, Oregon; Halifax, Nova Scotia; and Mazatlan, Sonora. At the continental scale, therefore, it may be better to protect a variety of patches of different habitat. Indeed, at the national level it is highly unlikely that the full wealth of a nation's biodiversity can be captured within a single reserve unless the reserve approaches the size of the entire country.

## Time Scales

Biological needs at all levels of complexity vary with time, as well, and there are intricate relationships between spatial and temporal processes. Reserves must be able to accommodate needs throughout the creature's life and the ecological processes that support it. Determining the appropriate scale for a reserve depends not only on the abundance and spatial ecology of species but on the flux of materials across landscapes as well. Frequency of events (a time-related variable) can be crucial. In other words, resources must be present when they are needed by species. The needs of individuals often vary seasonally in response to changing weather, habitat, and community characteristics. Nomadic and migratory terrestrial mammals (especially large ungulates and carnivores) may range over huge areas in the course of a year, for example, although occupying only a limited portion of that area at any one time. The species may require corridors to link populations and areas of seasonal use (Chapter 6). A small, isolated core reserve is likely to suffer a high rate of species loss because it may lack a resource when it is needed. Consider, for example, a reserve having only a single ephemeral (vernal) pond that is essential for the breeding of rare crustaceans or amphibians. During a drought, this pond may not contain sufficient water for reproduction of some species. If the drought persists, local populations of pond-dependent species would likely go extinct—particularly if the site could not be recolonized from afar. In this example, the time scale or temporal distributions of droughts must be a design consideration.

## Scale and the Distribution of Species

Some localities (tropical reefs, rain forests) have many species; others (deserts, polar areas) have few. In some regions, such as northern Canada, widely separated forest localities share virtually the same species; in other places, the composition of forest species changes dramatically over distances as short as a few meters (parts of Mexico). In some places, major barriers such as mountain ranges or oceans separate regions with virtually no overlap in species composition; in other regions, the composition of ecosystems differs little across oceans and mountain ranges. In some places there are no unique (endemic) species; in other places there are many endemic species.

These different phenomena have technical names: alpha, beta, and gamma diversity (Whittaker 1960, 1977; Sampson and Knopf 1982). Alpha- level biodiversity refers to the number of species that can be found within a small homogeneous area of a single habitat type; it is also called "richness." Hence alpha diversity constitutes a scale of species richness. Beta-level biodiversity refers to the rate of change (or degree of difference) in species composition across two or more habitat types (for example, forest to meadow) within the same general locality or watershed. Gamma-level biodiversity refers to the rate of change in species composition across large landscapes or major geographic discontinuities (for example, Utah to Kansas across the Rocky Mountains; Norway to Greenland across the North Atlantic; or Costa Rica to Venezuela across the Isthmus of Panama). Contrasting geographic patterns in the numbers and distributions of species are a matter of macroecology (Brown 1995; Brown and Maurer 1989). Macroecological patterns present special challenges for the protection of nature.

Beta and gamma diversity are measures of the rate of change in flora and fauna in space. Are these patterns an aspect of scale? Yes: like speedometer readings in a car, beta and gamma diversity are scales of *rates*. In a car, the speedometer reading is a good predictor of how much damage will occur if you hit an oncoming truck. The rates at which the flora and fauna change across the landscape are informative too; they guide us in decisions about the geographic distribution of reserves and about the chances of success or failure when adopting conservation strategies.

The significance of these patterns for biodiversity conservation is perhaps best illustrated with an example. Imagine three regions (I, II, III) with contrasting patterns of alpha, beta, and gamma diversity. Within each region, we sample four localities and find the patterns shown here:

| Region | I | II | III |
|---|---|---|---|
| Species in region | 6 | 6 | 3 |
| Alpha diversity (locality *mean*) | high (6) | low (3) | low (3) |
| Beta diversity | low | high | low |
| Gamma diversity | high | high | low |

Region I contains a total of six species; because all of them occur in the four localities, this region has high alpha and low beta diversity. In contrast, remote localities in Region I differ greatly in species composition. Region II harbors a total of six species, too, but local richness is low (three species per locality) and differences in composition between localities, whether near or far, are high. Region III shows low regional and local species richness (for three species, respectively), and differences in composition between adjacent and distant localities are low. In North America, places with diversity patterns similar to those described here could be: Region I, Central America; Region II, Mexico; Region III, Canada and most of the United States.

These differences in alpha, beta, and gamma diversity let us anticipate other considerations. Many macroecological studies of mammals have shown the existence of certain correlations between range and body size known for mammals. Based on these correlations, we can infer other characteristics of species in these regions as follows:

| Region | I | II | III |
|---|---|---|---|
| Average area of distribution | wide | narrow | wide |
| Average body mass | intermediate | small | large |
| Endemism | low | high | low |
| Similarity | high | low | high |

## Beta Diversity and Its Relevance

The average area of a species' distribution may be inversely correlated with beta diversity. Although no one has proved this relationship empirically, the relations can be intuitively derived from the preceding tables. Most mammal species in Central America are present in most localities, for example, similar to Region I. In Mexico, however, changes in the list of mammals from one site to the next are great, with most species having restricted distributional ranges (similar to Region II).

One factor behind the high beta diversity in areas such as Region II is often the high heterogeneity or physical differences among sites. Most species are restricted in their distribution, and many are endemic because of physical barriers; these barriers must be considered when designing reserve networks. The implication is that Mexico needs reserves in every region. This is because a group of localities with high beta diversity, such as Region II, show low values of similarity (because of the high turnover rate of species from one locality to the next) or high values of complementarity. A few reserves, even if very large, cannot protect Mexico's biodiversity. The same holds for other places with high rates of endemism, such as California, where many species have extremely localized distributions.

## The Relevance of Body Size, Home Range, and Trophic Status

Among mammals and birds, area of home range and body mass are positively correlated (McNab 1963; Schoener 1968). Everything else being equal, bigger animals need more room to forage. But everything is not equal. Carnivores need much more room than herbivores in general. Thus even though cougars (*Felis concolor*) and mule deer (*Odocoileus hervonius*) both weigh about 54 kilograms, carnivorous cougars have much larger home ranges than the herbivorous deer (Peters 1983). Average home range size during the breeding season for a male cougar is 29,300 hectares (Dixon 1983), while a mule deer's home range averages just 92 hectares (Mackie et al. 1983). As a consequence, regions with mostly small to medium-size herbivorous mammals and small carnivores (such as Australia) may not require such large reserves and extensive connectivity as those in other continents. But even where most mammals are relatively small, as in Mexico, small reserves may not be viable because of top-down regulation (Chapter 2) and because native carnivores, such as the large cats, depend on large areas for hunting and population viability.

### Endemism and Beta Diversity

Regions with high beta diversity often have a high proportion of endemic species. Indeed, Mexico shows a very high rate of endemism: about one-third of the mammal fauna are exclusive to the country. Therefore, it has been suggested that reserves in Mexico must be more numerous and more widely distributed than those in the United States or Canada in order to capture the same percentage of the area's biodiversity (Arita et al. 1997). Even in regions of high beta diversity and endemism, however, large reserves and connectivity among reserves will still be needed to maintain species or ecological processes with significant area requirements. In other words, it would be wrong to conclude that regions with high beta diversity—simply because most species have restricted distributions, small body size, and high resiliency—invariably need lower levels of connectivity among their protected areas than areas of low beta diversity. As noted earlier, areas of this type also harbor large, more mobile species that require connectivity for long-term population viability. In areas such as Region III, equivalent to most parts of the United States and Canada, connectivity among protected areas should be extensive in order to allow exchange among populations of large species—particularly migratory ungulates and carnivores that would otherwise not be viable.

In conclusion, the tools of macroecology provide new perspectives on the ecological and biogeographic variables relevant to the selection and design of networks of regional protected areas. The macroscopic view creates an ecological context in which to place individual reserves as well as reserve systems. Moreover, the macroscopic view of ecological systems makes it easier to identify transboundary issues of reserve design and management—not always possible when planning is restricted to traditional scales.

## Disturbance and Scale

With regard to disturbance events, both natural and anthropogenic, spatial and temporal scale should be considered. Frequency is an important temporal scale factor. If fire frequency is artificially high, it can cause the local extirpation of plant species (Bradstock et al. 1996). If it is too low, as happens when fires are suppressed, it virtually guarantees that when fires finally do occur they will be major conflagrations. Frequency and seasonality can be used to our advantage; burning at the right season is frequently used to control invasive exotics.

Disturbance is often classified according to severity, which has several components. Considering fire, for example, these components include intensity (how hot), season, and size (relative and absolute). Thus a fire's impact on a protected area depends in large part on its size and the temperature. A large, hot fire in the summer that kills every tree in a forest is not the same as a cool fire in the late fall that kills only saplings. Pickett and Thompson (1978) addressed the issue of disturbance with their concept of a "minimum dynamic area"—that is, "the smallest area with a natural disturbance regime which maintains internal recolonization sources, and hence minimizes extinction." Different disturbance regimes create patches over a range of sizes, each with a unique duration. Protected areas designed to persist in the face of a variety of disturbance events, covering a range of temporal scales, must be of a size and position on the landscape that can accommodate these events while ensuring the survival of affected species.

Frequency of fire and other disturbance events is also important with regard to plant succession, soil formation, animal and plant movements, population changes, and other ecological processes (Pickett and Thompson 1978; Turner et al. 1993). And for natural disturbance events, recovery is an integral part of the process. For example, a forest begins recovering almost immediately following a fire (Elfring 1989; Christensen et al. 1989); the recovery period, however, depends on the size and intensity of the disturbance. Examples of natural and anthropogenic disturbance events are presented in Table 2.1.

## Table 2.1. Examples of Natural and Anthropogenic Disturbances *

| Natural Events | Anthropogenic Events |
| --- | --- |
| fire | residential development |
| disease epidemic | road, trail, railroad line |
| flood | telephone line, electrical power line |
| drought | dam, water diversion, canal |
| hurricane/tornado/windstorm | commercial development |
| avalanche/landslide | modern agriculture |
| volcanic eruption | mining |
| ice storm | logging |
|  | grazing |

* Entries in italics connote reversible disturbances, while those in roman usually represent long-term or permanent conversion of habitat.

The relationships between the spatial scale of disturbance, the frequency of disturbance, and the length of time required for recovery determine the rate of recovery. Some habitats require periodic fire to maintain all their natural components of biodiversity: Ponderosa pine forests (*Pinus ponderosa*), New Jersey pine barrens, peninsular Michigan jack pine (*Pinus banksiana*) barrens, California chaparral and sage scrub, Florida longleaf pine (*Pinus palustris*)/wiregrass (*Aristida stricta*). In these cases, managers usually try to mimic the natural, spatial scale and temporal frequency of burning. In turn, these depend on the natural history of the dominant species, their density, and their generation time.

Preliminary analysis suggests that the recovery time from disturbance scales compares roughly with the square root of the area disturbed (Dobson et al. 1997). Furthermore, the rates of recovery of damaged ecosystems appear to be generally similar for natural and anthropogenic disturbances of similar magnitude. Clearly it is important to gather more empirical data about this purported relationship. Above all, we need to know if response rates vary among different ecosystems (which seems likely) and depend on the type and severity of disturbance (which is less certain). Moreover, a full understanding of disturbance requires information on the frequency at which disturbance events of different sizes occur. Although disturbances such as massive meteor strikes obviously occur at a lower frequency than localized disasters such as lightning strikes or tree falls, the relationship between frequency, magnitude of disturbance, and recovery time may give us insight into the resiliency of different systems to disturbance.

## The Scale of Interactions Between Ecosystems

There is increasing evidence for interactions between what might otherwise appear to be spatially and functionally distinct ecosystems. These interactions have major implications for the design of reserves. The open sea, for instance, is known to influence coastal reefs by transporting the larval life-history stages of numerous plant, invertebrate, and fish species (Roughgarden et al. 1988; Underwood and Fairweather 1989). Important interactions are known or suspected to occur between sea and land as well. Such effects may be particularly important when there is a strong disparity in some limiting resource between the two ecosystems—in which case the resource-rich ecosystem may subsidize the one that is resource-poor. Gary Polis and colleagues have demonstrated this effect on desert islands in the Gulf of California where coastal terrestrial ecosystems are strongly subsidized by marine production (Polis and Hurd 1995, 1996; Polis et al. 1997; Anderson and Polis 1998).

Similar effects may occur where migratory or nomadic fishes transport vast quantities of biomass from high-production to low-production environments. In the tropics, where the majority of diadromous (living in both fresh and salt water) fish species are catadromous, living in freshwater, spawning in the sea (Gross et al. 1988), net material and nutrient flux from land to sea is likely because of fish movement. At high latitudes (Gross et al. 1988) the preponderance of anadromous species (such as salmon) would be expected to reverse this pattern. Similar transport of nutrients from the sea to upland habitats in the tropical Pacific by seabirds has been postulated (Steadman 1997).

While strong evidence is still lacking, the once vast numbers of Pacific salmon may have been a major player in the ecology of the boreal forests and freshwater streams and rivers of western North America (Willson et al. 1998; Bilby et al. 1996). The striking dichotomies of population density and home range size between coastal versus inland populations of brown bears (*Ursus arctos*) are undoubtedly influenced to some degree by this effect (Miller et al. 1997). There is also reason to suspect that seabird populations, which roost and breed on land while feeding at sea, may subsidize certain terrestrial ecosystems by delivering nutrients in the form of their carcasses, feces, and discarded food. This effect may be especially strong at high latitudes, where soils are often nutrient-impoverished while the adjacent oceanic environment is nutrient rich and highly productive.

An important class of spatial subsidy involves aeolian ecosystems (Swan 1963) in which wind-borne nutrients, trace elements, and detritus fuel production over tremendous distances. One example of this phenomenon involves the tropical forests of Amazonia where soils are intrinsically nutrient-impoverished. Production in this system is thought to be regulated by the intercontinental transport of phosphorus-rich soil in the form of dust blown from the deserts of Africa (Swap et al. 1992). Thus a change in climate—such as increased rainfall in Africa—could strongly influence the function of rain-forest ecosystems in the Amazon Basin. Additional examples of spatially subsidized nutrient and food links could be cited, and other cases almost certainly are yet to be discovered.

These examples of flows between habitats or vegetation types should stand as a warning to those designing reserves. One should not assume that geographically distinct ecosystems are functionally distinct or self-sustaining. When planning reserves, one must consider the species and processes that transport energy across ecosystems and provide for their continued viability. The disruption of connectivity between "unconnected" ecosystems may result in diminished ecological processes and weakened ecosystem integrity. It may sound unscientific, but the old ideas about the unity of the biosphere are gaining in respectability.

The processes that mediate the functioning of ecosystems operate on a variety of spatial scales. Viable populations of keystone species that regulate species diversity (Chapter 2) are likely to require more space than processes such as nutrient cycling and pollination. Pollination by insects of an individual plant takes place on a scale of centimeters to meters, for example, whereas the space required to maintain viable populations of pollinator (and plant) populations may be on the order of 100 to 1000 square kilometers. Similarly, the search, pursuit, and capture interactions between large predators and their prey occur at spatial scales of 100 square meters to 1 square kilometer, yet viable populations of carnivores may require areas on the order of 25,000 square kilometers or larger (U.S. Fish and Wildlife Service 1992).

One approach to determining the minimum size required for an effectively functioning ecosystem is to ascertain the reserve size below which major ecosystem processes fail—though this is a difficult task. Alternatively, the incidence functions (Diamond 1975) of species that mediate key ecosystem processes could be empirically determined and used to define reserve size thresholds. But the beguiling simplicity of these approaches is undermined by the fact that in different habitats, similar ecological processes are mediated by species with different area requirements. For example, salmon, grizzly bears, and bald eagles all may play a major role in "up-slope" nutrient recycling and calorie transport in the coastal forests of North America (Hilderbrand et al. 1996; Willson et al. 1998). These processes take place on the spatial scale of individual watersheds (100–1000 square kilometers).

The recycling of nutrients in East African grasslands, by contrast, may occur on a much smaller geographic scale. Over the course of one or two years, patches of grassland become preferred grazing and defecation locations by ungulates, thus becoming nutrient-rich (McNaughton et al. 1988; McNaughton 1988). The areas of nutrient enrichment, which tend to be on the order of 1 square kilometer or less, are smaller than the area over which nutrients are removed (10–50 kilometers). In some places, these preferred patches reflect underlying concentrations of certain trace elements. Other, more ephemeral, grazing sites simply reflect behavioral choices of the ungulates—perhaps as a response to decreased local predation pressure. Some of these ungulates and omnivores may simultaneously facilitate other ecosystem processes, such as seed dispersal. Deposition of seed in nutrient-rich patches may enhance seedling survival. Similar processes occur at a still finer spatial scale in desert habitats, where ants collect seeds from an area (100-meter radius) around their nests and discard unused parts of these seeds next to the nest entrance (about 1 square meter). This process results in a mosaic of patches in the local plant community that are apparent over much larger spatial scales (10–100 square kilometers). All these examples underscore the importance of determining the most area-sensitive process in each habitat or ecosystem type and ensuring that the area preserved exceeds the area needed to maintain essential flows.

## Learning to Think Like Mountains and Continents

Increasing interest in the conservation of biodiversity has created increasing awareness of scale-related issues. Above all, this goal has required that attention be given to ecological processes operating over large areas and long periods of time. Scientific research continues to reveal not just the interdependence of species within communities, but also profoundly important interactions between ecosystems and ecosystem processes, sometimes widely separated in time and space. The job of conserving biodiversity requires broadening the scope of temporal and spatial considerations beyond the traditional focus of resource management. If the goal is preservation of existing biota and its long-term evolutionary potential, this broad-scale, long-term framework is absolutely essential.

In this chapter we have highlighted several key principles. For example: scale-related issues are asymmetrical. While we may have some chance of preserving field mice if we provide for the needs of grizzly bears, we have virtually no chance of preserving bears if we plan only for mice. This simple principle is the premise behind building conservation programs around surrogates such as single species, species richness, vegetation types, flagships, umbrellas, keystones, or enduring physical structures. While many more studies are needed, such approaches are commonplace because such simplifications match the monetary and analytical resources of most conservation programs.

Another principle is that the spatial aspects of reserve design are governed largely by the wide-ranging or otherwise most mobile elements of the biotic community or abiotic environment. Thus carnivores and migratory animals are disproportionately important in conservation programs—even those focused on the preservation of biological diversity (Chapter 3). Communities also have self-organizing features that often operate at orders of magnitude greater than the scale at which the affected organisms and processes operate. The size, frequency, and intensity of disturbances are also critical considerations in conservation design. Populations and species will persist only where there is enough contiguous occupied habitat to ensure that local declines in productivity or survival caused by disturbances will not lead to local extirpations. Collectively these emergent principles constitute an imperative for designing reserves—or systems of reserves—much larger than most that exist today.

Creating a system of reserves populated by the full range of biological diversity and stretching from Panama to the Arctic Circle is a bold vision. If it is to be realized, we must begin to think "like a mountain," as Aldo Leopold advised us—that is, we must think on spatial and temporal scales that we rarely consider in our daily lives. And if we are to maintain the evolutionary potential of species that constitute these systems, we must learn to think in terms of all the relevant scales: spatial, temporal, levels of organization, species richness, rates of species turnover, and rates of disturbance (Wiens 1989; Wiens et al. 1986; Franklin 1993; Orians 1993).

# References

Allen, T. F. H., and T. B. Starr. 1982. *Hierarchy perspectives for ecological complexity.* Chicago: University of Chicago Press.

Anderson, W. B., and G. A. Polis. 1998. Marine subsidies of island communities in the Gulf of California: Evidence from stable carbon and nitrogen isotopes. *Oikos* 81:75–80.

Arita, H. T., F. Figueroa, A. Frisch, P. Rodriquez, and K. Santos-del Prado. 1997. Geographical range size and the conservation of Mexican mammals. *Conservation Biology* 11:92–100.

Beier, P. 1993. Determining minimum habitat areas and habitat corridors for cougars. *Conservation Biology* 7:94–108.

Bilby, R. E., B. R. Fransen, and P. A. Bisson. 1996. Incorporation of nitrogen and carbon from spawning coho salmon into the trophic system of small streams: Evidence from stable isotopes. *Canadian Fisheries and Aquatic Sciences* 53:164–173.

Bradstock, R. A., M. Bedward, J. Scott, and D. A. Keith. 1996. Simulation of the effect of spatial and temporal variation in fire regimes of the population viability of a Banksia species. *Conservation Biology* 10:776–784.

Brown, J. H. 1995. *Macroecology.* Chicago: University of Chicago Press.

Brown, J. H., and B. A. Maurer. 1989. Macroecology: The division of food and space among species on continents. *Science* 243:1145–1150.

Buchmann, S. L., and G. P. Nabhan. 1996. *The forgotten pollinators.* Washington, D.C.: Island Press.

Caicco, S. L., J. M. Scott, B. Butterfield, and B. Csuti. 1995. A gap analysis management status of the vegetation of Idaho. *Conservation Biology* 9:498–511.

Christensen, N. L., J. K. Agee, P. F. Brussard, J. Hughes, D. H. Knight, G. W. Min-shall, J. M. Peek, S. J. Pyne, F. J. Swanson, J. W. Thomas, S. Wells, S. E. Williams, and H. A. Wright. 1989. Interpreting the Yellowstone fires of 1988. *BioScience* 39(10):678–685.

Davis, F. W., P. A. Stine, D. M. Stoms, M. I. Borchert, and A. D. Hollander. 1995. Gap analysis of the actual vegetation of California: 1. The southwest region. *MadroZo* 42:40–78.

Diamond, J. D. 1975. The island dilemma: Lessons of modern biogeographic studies for the design of natural preserves. *Biological Conservation* 7:129–146.

Dixon, K. R. 1983. Mountain lion. In J. A. Chapman and G. A. Feldhamer (eds.), *Wild mammals of North America*. Baltimore: Johns Hopkins University Press.

Dobson, A. P., A. D. Bradshaw, and A. J. M. Baker. 1997. Hopes for the future: Restoration ecology and conservation biology. *Science* 277:515–521.

Elfring, C. 1989. Yellowstone: Firestorm over fire management. *BioScience* 39(10):667–672.

Frankel, O. H., and M. E. Soulé. 1981. *Conservation and evolution*. Cambridge: Cambridge University Press.

Franklin, J. F. 1993. Preserving biodiversity: Species, ecosystems, or landscapes. *Ecological Applications* 3:202–205.

Gilpin, M., and I. Hanski. 1991. *Metapopulation dynamics: Empirical and theoretical considerations*. London: Academic Press.

Gross, M. E., R. M. Coleman, and R. M. McDowall. 1988. Aquatic productivity and the evolution of diadromous fish migrations. *Science* 239:1291–1293.

Higgs, A. J., and M. B. Usher. 1980. Should nature reserves be large or small? *Nature* 285:568–569.

Hilderband, G.V., S. D. Farley, C. T. Robbins, T. A. Hanley, K. Titus, and C. Servheen. 1996. Use of isotopes to determine diets of living and extinct bears. *Canadian Journal of Zoology* 74:2080–2088.

Hunter, M. L., G. L. Jacobson, and T. Webb. 1988. Paleoecology and the coarse-filter approach to maintaining biological diversity. *Conservation Biology* 2:375–385.

Jennings, M. J., and J. M. Scott. 1993. Building a macroscope: How well do places managed for biodiversity match reality? *Renewable Resources Journal,* Summer: 16–20.

Kareiva, P. M., and M. Anderson. 1988. Spatial aspects of species interactions: The wedding of models and experiments. In A. Hastings (ed.), *Community ecology*. New York: Springer-Verlag.

Mackie, R. J., K. L. Hamlin, and D. F. Pac. 1983. Mule deer. In J. A. Chapman and G. A. Feldhamer (eds.), *Wild mammals of North America*. Baltimore: Johns Hopkins University Press.

McNab, B. K. 1963. Bioenergetics and the determination of home range size. *American Naturalist* 97:133–140.

McNaughton, S. J. 1988. Mineral nutrition and seasonal migration of African migratory ungulates. *Nature* 334:343–345.

McNaughton, S. J., R. W. Ruess, D. A. Frank, and S. W. Seagle. 1988. Large mammals and process dynamics in African ecosystems. *BioScience* 38:794–800.

Meffe, G. K., and C. R. Carroll. 1997. *Principles of conservation biology.* 2nd ed. Sunderland, Mass.: Sinauer.

Miller, B., R. Reading, J. Stritthold, C. Carroll, R. Noss, M. E. Soulé, O. Sanchez, J. Terborgh, and D. Foreman. 1999. Focal species in the design of reserve networks. *Wild Earth* 8(4):81–92.

Miller, S. D., G. C. White, R. A. Sellers, H. V. Reynolds, J. W. Schoen, K. Titus, V. G. Barnes, R. R. Nelson, W. B. Ballard, and C. C. Schwartz. 1997. Brown and black bear density in Alaska using radiotelemetry and replicated mark-resight techniques. *Wildlife Monographs* 133.

Newmark, W. D. 1987. Mammalian extinctions in western North American parks: A landbridge perspective. *Nature* 325:430–432.

———. 1995. Extinction of mammal populations in western North American national parks. *Conservation Biology* 9:512–526.

Noss, R. F. 1990. Indicators for monitoring biodiversity: A hierarchical approach. *Conservation Biology* 4:355–364.

Noss, R. F., E. T. LaRoe III, and J. M. Scott. 1995. *Endangered ecosystems of the United States: A preliminary assessment of loss and degradation.* Biological report 28. Washington, D.C.: National Biological Service.

O'Neill, R. V., D. L. DeAgelis, J. B. Waide, and T. F. H. Allen. 1986. *A hierarchical concept of ecosystems.* Princeton: Princeton University Press.

Orians, G. H. 1993. Endangered at what level? *Ecological Applications* 3:206–208.

Peters, R. H. 1983. *The ecological implications of body size.* Cambridge studies in ecology. New York: Cambridge University Press.

Pickett, S. T. A., and J. N. Thompson. 1978. Patch dynamics and the design of nature reserves. *Biological Conservation* 13:27–37.

Polis, G. A., and S. D. Hurd. 1995. Extraordinarily high spider densities on islands: Flow of energy from the marine to terrestrial food webs and the absence of predation. *Proceedings of the National Academy of Sciences* 92:4382–4386.

———. 1996. Linking marine and terrestrial food webs: Allochthonous input from the ocean supports high secondary productivity on small islands and coastal land communities. *American Naturalist* 147:396–423.

Polis, G. A., W. B. Anderson, and R. D. Holt. 1997. Towards an integration of landscape and food web ecology: The dynamics of spatially subsidized food webs. *Annual Review of Ecology and Systematics* 28:289–316.

Power, M. E., D. Tilman, J. A. Estes, B. A. Menge, W. J. Bond, L. S. Mills, G. Daily, J. C. Castilla, J. Lubchenco, and R. T. Paine. 1996. Challenges in the quest for keystones. *BioScience* 46:609–620.

Roughgarden, J., S. Gaines, and H. Possingham. 1988. Recruitment dynamics in complex life cycles. *Science* 241:1460–1466.

Sampson, F., and F. L. Knopf. 1982. In search of diversity ethics for wildlife management. *Transactions of the North American Wildlife and Natural Resources Conference* 47:421–431.

Schoener, J. W. 1968. Sizes of feeding territories among birds. *Ecology* 49:123–141.

Scott, J. M., and B. Csuti. 1997. Noah worked two jobs. *Conservation Biology* 11:1255–1257.

Scott, J. M., F. Davis, B. Csuti, R. Noss, B. Butterfield, C. Groves, H. Anderson, S. Caicco, F. D. D'Erchia, T. C. Edwards Jr., J. Ulliman, and R. G. Wright. 1993. *Gap analysis: A geographic approach to protection of biological diversity.* Wildlife monograph 123.

Shelford, V. E. 1933. Ecological Society of America: A nature sanctuary plan unanimously adopted by the society. *Ecology* 14:240–245.

Simberloff, D. S. 1986. Design of nature reserves. In M. B. Usher (ed.), *Wildlife Conservation Evaluation.* London: Chapman & Hall.

Simberloff, D. S., and G. G. Abele. 1976a. Island biogeography theory and conservation practice. *Science* 191:285–286.

———. 1976b. Island biogeography and conservation: Strategy and limitations. *Science* 193:1032.

Soulé, M. E., and R. F. Noss. 1998. Rewilding and biodiversity conservation: Complementary goals for continental conservation. *Wild Earth* 8(3):18–28.

Soulé, M. E., and D. Simberloff. 1986. What do genetics and ecology tell us about the design of nature reserves? *Biological Conservation* 35:9–40.

Steadman, D. W. 1997. Extinctions of Polynesian birds: Reciprocal impacts of birds and people. In P. V. Krich and T. L. Hunt (eds.), *Historical ecology in the Pacific Islands: Prehistoric environmental and landscape change.* New Haven: Yale University Press.

Swan, L. W. 1963. Aeolian zone. *Science* 140:77–78.

Swap, R., M. Garstang, S. Greco, R. Talbot, and P. Kallberg. 1992. Saharan dust in the Amazon Basin. *Tellus Series B, Chemical and Physical Meteorology* 44:133–149.

Turner, M. G., W. H. Romme, R. H. Gardner, R. V. O'Neill, and T. D. K. Krauts. 1993. A revised concept of landscape equilibrium: Disturbance and stability on scaled landscapes. *Landscape Ecology* 8:213–227.

Underwood, A. J., and P. G. Fairweather. 1989. Supply-side ecology and benthic marine assemblages. *Trends in Ecology and Evolution* 4:16–20.

U.S. Fish and Wildlife Service. 1992. *Revised recovery plan for the eastern timber wolf.* St. Paul: U.S. Fish and Wildlife Service.

U.S. Office of Technology and Assessment. 1987. *Technologies to maintain biodiversity.* OTA-F-330. Washington, D.C.: Government Printing Office.

Whittaker, R. H. 1960. Vegetation of the Siskiyou Mountains, Oregon and California. *Ecological Monographs* 30:279–338.

———. 1977. Species diversity in land communities. *Evolutionary Biology* 10:1–67.

Wiens, J. A. 1989. Spatial scaling in ecology. *Functional Ecology* 3:385–397.

Wiens, J. A., Addicott, J. F. Case, and J. Diamond. 1986. Overview: The importance of spatial and temporal scale in ecological investigations. In J. Diamond and T. J. Case (eds.), *Community Ecology.* New York: Harper & Row.

Willson, M. F., S. M. Gende, and B. H. Marston. 1998. Fishes and forests. *BioScience* 48:455–462.

# 3

# The Role of Top Carnivores in Regulating Terrestrial Ecosystems

*John Terborgh, James A. Estes, Paul Paquet,
Katherine Ralls, Diane Boyd-Heger, Brian J.
Miller, and Reed F. Noss*

The vast majority of species inhabiting the earth today have existed for more than a million years (Stanley 1987; May et al. 1995). Significantly, the last million years have been, climatically, among the most turbulent of the last 500 million years, with major and often abrupt changes in mean temperature, rainfall, glaciation, sea level, and extent of sea ice (Pielou 1991). Notwithstanding the extraordinary climatic instability of the recent past, extinction rates have not been particularly high (Coope 1995). In the absence of human beings, therefore, most plant and animal species are remarkably resilient to natural environmental instabilities of the kinds that prevailed over the Pleistocene era. How can we account for this resilience to extinction of wild species? If we knew the answer, it would be of immeasurable help in reducing the rate of extinction in our own time. Extinction rates are acknowledged to be hundreds or thousands of times higher today than they were in the prehuman past (Wilson 1992; May et al. 1995; Ehrlich 1995). Scores of studies have asked why a particular species or population went extinct or became endangered. In some cases—as in the overharvest of the dodo and great auk (Diamond 1982)—the cause is obvious. But in many others, it is hard to distinguish proximate from ultimate causes (Caughley 1994).

Both physical and biological processes are important in preserving biodiversity. An appropriate disturbance regime, for example, is considered essential to maintaining diversity in plant communities (Connell 1978). In a variant on the same theme, natural grasslands often depend on herbivores for opening sites that help plants colonize, but today livestock have widely replaced native herbivores—often with devastating impacts on plant communities. Predation can play an analogous role in reducing inter- and intraspecific competition for resources among prey species. Simple predator/prey models describe feedback processes leading to a stable point or stable limit cycle, in which the numbers of predators and prey come to equilibrium or oscillate within circumscribed limits. But widespread elimination of top predators from terrestrial ecosystems the world over has disrupted the feedback process through which predators and prey mutually regulate each other's numbers.

In this chapter we focus on predation as a key process in the natural maintenance of biodiversity. The role of predation has become a matter of intense interest to conservationists because mounting evidence, as we shall see, points to the pivotal role of predation in helping to preserve the biodiversity of terrestrial communities. On every continent, top predators are now restricted to tiny fractions of their former ranges so that the integrity of biological communities over large portions of the earth's terrestrial realm is threatened by grossly distorted predation regimes. Even where they are present, population densities of top predators tend to be so low, and their behavior so secretive, that sightings are infrequent. Most biologists prefer to study species that are common, small, and easily manipulated. Many academics dismiss field studies of large carnivores as "unscientific" because sample sizes are typically small and controlled experimentation difficult. Carnivore biology has thus been left to a small coterie of hardy devotees whose work, if not ignored, lies well outside the mainstream. The role that top predators play in terrestrial ecosystems, therefore, remains ill defined and contentious. (See Erlinge et al. 1984, 1988; Kidd and Lewis 1987; Terborgh 1988; Hunter and Price 1992; Power 1992; Strong 1992; Wright et al. 1994; Estes 1996.) At the end of a literature review, for example, Polis and Strong (1996) conclude "that trophic cascades and top-down community regulation as envisioned by trophic-level theories are relatively uncommon in nature." Here, after reviewing an overlapping body of literature, we come to the opposite conclusion.

Whether contentious or not, it is crucial to define the role of top predators because the stakes are enormous. If, as we conclude here, top predators are often essential to the integrity of ecological communities, it will be imperative to retain top predators or restore them to as many parts of the North American continent as practical. Failure to do so will result in distorted ecological interactions that, in the long run, will jeopardize biodiversity.

## Theory

What is at issue in the current debate over "top-down" versus "bottom-up" processes (Matson and Hunter 1992)? "Top-down" means that species occupying the highest trophic level (top carnivores) exert a controlling influence on species at the next lower level (their prey) and so forth down the trophic ladder. The definition can be made operational in a thought experiment. Under top-down regulation, the removal of a top predator (or better, the entire guild of top predators) results in an appreciable population increase in the prey. It is thereby demonstrated that productivity (the food supply available to the prey) was not the proximal factor limiting prey numbers. Conversely, if removal of the guild of top predators does not lead to increases in the numbers of prey, we must conclude that the prey were proximally limited by something else—most likely the food supply.

We can ask parallel questions about the bottom rung on the trophic ladder. Suppose we could increase the long-term productivity of an ecosystem experimentally—let us say by adding water to a desert or nutrients to a barrens (Wedin and Tilman 1993). If the increase in plant growth resulting from the artificial input then led to an increase in the biomass of consumers (herbivores such as rabbits and deer), we could conclude that the consumers were under bottom-up control. If we found no increase in consumer biomass, this would imply that something other than productivity was limiting—plant antiherbivore defenses, or predators, to mention two possibilities (Oksanen 1983). Even by the admittedly simple operational criteria just presented, it should be evident that top-down versus bottom-up is not merely an either/or proposition. If we could add water or fertilizer to an ecosystem, the number of consumers could increase even in the presence of predators—implying bottom-up regulation. Simultaneously, say, in a different experimental plot, consumers could increase in response to predator removal without external inputs such as water or fertilizer—implying top-down regulation (Brett and Goldman 1997).

Both top-down and bottom-up regulation can operate concurrently in the same system. In the presence of predators, herbivores are secretive and act as time-minimizers, thereby maximizing their survival. That is, they endeavor to spend as little time feeding (when they are exposed to predators) as possible. Most of the time is spent in secure places—in burrows or dense thickets, for example, or in naturally protected spots such as steep mountain slopes or ledges (bighorn sheep and mountain goats). If predators are removed, then the quest for security ceases to be the leading regulator of prey behavior; now consumers are free to feed when and where they want, becoming energy-maximizers, thereby maximizing fecundity. The switch in prey behavior from time-minimizer to energy-maximizer in response to differing levels of perceived predator threat introduces complexity into the system and allows both top-down and bottom-up regulation to operate simultaneously or to varying degrees. (See Power 1992; Werner and Hall 1988; Abrams 1993; Werner and Anholt 1993; Englund 1997.) Another layer of complexity is added by herbivory-induced plant defenses. Damage to foliage can stimulate plants to increase levels of herbivore-deterring chemicals in their tissues— thereby reducing the food supply available to herbivores (a bottom-up effect). It is the extraordinary complexity of trophic interactions that has made the issue of top-down versus bottom-up a matter of so much contention among ecologists.

Top-down effects have been shown to act on communities in two fundamentally different ways. One is through preferential feeding on a prey species that, in the absence of predation, is capable of competitively excluding other species that depend on a limiting resource. Thus, over an intermediate range of predation intensities, species diversity in the prey guild is enhanced over that which occurs in the overabundance or absence of predators. Here we refer to this process as the "Paine effect." A more generalized form of this process, known as the intermediate disturbance model of species diversity, has been demonstrated in a variety of systems (Connell 1978, Sousa 1984).

The second way in which predators influence their communities is through a cascade of interactions extending through successively lower trophic levels to autotrophs at the base of the food web (Carpenter and Kitchel 1993). In trophic cascades, the autotrophs are either enhanced by reduced herbivory or limited by increased herbivory, depending on whether the number of trophic levels is odd or even (Power 1992). The top-down model predicts that each trophic level is potentially limited by the next level up. For intact three-level systems, therefore, predators limit herbivores, thus releasing producers from limitation by herbivory. Since there is little unambiguous evidence from terrestrial systems for trophic cascades involving three or more levels, a number of studies have looked only at component steps—for instance, evidence that herbivores limit plants when the predators are missing and evidence that herbivores are limited by predators (Estes 1996).

## Empirical Foundations: The Paine Effect

If terrestrial carnivores were not so inherently difficult to study, we might have understood their roles long ago. The simpler conditions character-istic of certain aquatic systems have facilitated investigation, however, and the keystone role of predators is now established beyond dispute. Paine (1966) was the first to provide incontrovertible evidence. By remov-ing the predatory starfish *Pisaster ochraceous* from sections of the inter-tidal zone of the rocky Washington coastline, he showed that the diver-sity of the attached invertebrates subsequently declined as a superior competitor, the mussel *Mytilus californicus,* gradually occupied all avail-able space, thereby excluding other species from the community. It is important to note that *Mytilus* is the preferred prey of *Pisaster,* so that the action of the predator is selective removal of the dominant competitor—an act that exposes attachment sites that can be exploited by other species. Further studies of sessile intertidal communities have amply sup-ported Paine's result (with some geographical variation and local excep-tions). The effect of a top predator is reduced, for example, when it does not feed preferentially on the dominant competitor among the potential prey species (Menge 1992; Menge et al. 1994; Menge 1995). The primary effect of a top predator in the intertidal system is thus seen in regulating the diversity of the prey community. This is the Paine effect.

The presence/absence of a predator influences the productivity and biomass of the intertidal prey community because space (attachment sites) is the limiting resource. The productivity that supports the inter-tidal community is almost entirely imported from the open ocean—an example of a spatially subsidized food web. Interactive links between ses-sile intertidal predators and the productivity of the system are thus weak to nonexistent. Terrestrial and aquatic systems involving mobile organ-isms may show different dynamics, however, because consumers and predators are free to come and go and many of the component species have long lifetimes. And unlike Paine's rocky intertidal system, which can be studied on the scale of a few square meters, terrestrial and open-water aquatic systems must be studied on vastly larger spatial scales because the important predators and consumers may have low population densities and range over large areas. These daunting obstacles to the careful analy-sis of mobile predator/prey systems have been major impediments to scientific progress. Now, with results emerging from some long-term studies and the first large-scale predator-exclusion experiments, the time is ripe for a synthesis.

## Anecdotal Evidence

In the hope of arriving at some general conclusions, we now review evidence relevant to understanding the role of top carnivores. Our emphasis is on terrestrial ecosystems and large vertebrates, especially mammals. Although open aquatic systems provide many parallels, they are mentioned here only briefly. The evidence can be broadly categorized as anecdotal or experimental, though the dividing line between the two categories is not always distinct. Here we refer to evidence derived from natural perturbations and experiments lacking controls as "anecdotal."

HERBIVORE RELEASE ONTO PREDATOR-FREE ISLANDS. Sailors of yore introduced herbivores to predator-free islands throughout the Seven Seas to ensure themselves of a supply of meat on subsequent voyages. Horses, cattle, caribou, sheep, goats, pigs, and rabbits are among the animals introduced, singly or in combinations, to countless islands around the world (Carlquist 1974; Bramwell 1979; Coblentz 1978, 1990; Crosby 1986; Vitousek 1988). Few of these introductions were carefully monitored, so they can hardly be considered scientific studies. Nevertheless, in numerous instances (Ascension, Aldabra, Juan Fernández, California Channel Islands, St. Mathews Island, St. George Island) the introduced herbivores increased without check until they devastated the native vegetation of the island—at which point populations of the herbivores themselves often crashed (Klein 1968; Carlquist 1974; Coblentz 1980; Cronk 1980).

Destruction of the vegetation of predator-free islands by herbivores is unambiguously a top-down effect. Herbivores do not ordinarily destroy the vegetation of large landmasses supporting top predators, so it is tempting to attribute their massive impacts on islands to the absence of predators (Hairston et al. 1960). Another interpretation is possible, however, so the conclusion of top-down regulation is not the only one that can be drawn. The vegetation of islands lacking native vertebrate herbivores must experience relaxed selection for antiherbivore defenses and hence might be exceptionally vulnerable to introduced herbivores (Carlquist 1974; Bowen and van Vuren 1997). Without additional information, we cannot distinguish the two interpretations; but under the right circumstances, both may be correct.

PREDATOR ELIMINATION. Humans have eliminated top predators over much of the globe, drastically reducing the geographical ranges of many species, including wolves, bears, tigers, lions, and many less intimidating beasts. Nevertheless, herbivores have generally not overrun predator-free portions of the planet, as we would expect if herbivore populations were indeed under top-down control. The reason in this case appears obvious. Large vertebrate herbivores are also the prey of human beings, and in many places they have been reduced to low densities or extirpated by human overhunting (Redford 1992). In many regions, introduced livestock substitute for missing native ungulates. Untangling the effects of predator removal from those of hunting and introduced livestock is an almost impossible task in most situations.

One common, nonexperimental situation that conforms to the requirements of a proper test of top-down control is increasingly attracting scientific attention. It is found in suburban areas and parklands in the United States from which top predators were eliminated long ago and where hunting is now prohibited. Mammals that would have been part of the prey pool of missing carnivores such as wolves and cougars have, despite high rates of roadkill, become notoriously abundant to the point that some of them are now nuisances: by being road hazards (deer, moose); by browsing ornamental shrubbery (deer); by raiding trash cans (opossums, raccoons); by preying on birds (house cats) and their nests (cats, raccoons); by destroying vegetable gardens (deer, woodchucks, ground squirrels), and by flooding people's yards (beaver; Garrott et al. 1993). The problem of mammalian overabundance in predator-free portions of North America has become so widespread and so severe that it was recently the topic of a major symposium hosted by the Smithsonian Institution (McShea et al. 1997).

If top-down processes, as elucidated by Paine, are important in terrestrial ecosystems, then the removal of top predators must lead to reduced diversity in the next lower trophic level. The obvious experiment to test this proposition was preempted long ago, however, by megafaunal overkill. What is now the eastern United States once supported an impressive galaxy of large herbivores—including elephants, tapirs, ground sloths, capybaras, giant beaver, and others—but today it supports only one or two, the white-tailed deer and moose. Certainly white-tailed deer, raccoons, woodchucks, and beaver have proliferated dramatically in the absence of large carnivores, but it seems highly unlikely that any of these animals could ever drive another to extinction via exploitation competition (depletion of the food supply). Are we to conclude, then, that the Paine mechanism is inoperative on land?

This conclusion is not inevitable. The Paine effect operates through the monopolization of space, not resource competition. The few examples from terrestrial ecosystems that resemble a Paine effect involve small rodents. Small island communities of native rodents are conspicuously vulnerable to invasion and monopolization by a behaviorally dominant species. Small, eighty-year-old islands in Lake Gatun, Panama, are today occupied only by the spiny rat, *Proechimys semispinosus,* even though central Panamanian forests support sixteen species of rodents, at least some of which were presumably present on these islands at isolation (Adler and Seamon 1991). Other examples emanate from predator-free islands where introduced rats, particularly *Rattus rattus,* or mice have replaced other rodent species (Brosset 1963; Berry and Tricker 1969; Lynam 1997). Even on the large landmass of Madagascar, where a wide complement of predators is present, there is mounting evidence that introduced *Rattus* is displacing native rodents (Goodman 1995). Such competitive displacements of several species by one are not true Paine effects, because space is not limiting, but like the Paine mechanism they do occur in the absence of normal predation.

Although biologists have not fully documented the exact mechanism by which a single rodent species can, in the absence of predators, replace a community of other species, some rat species (such as *Rattus rattus*) are aggressive toward other rodents and are known to attack their nests and kill the young. If overt aggression is involved, then the takeover of predator-free islands by an aggressive rodent species would involve a form of spatial monopolization analogous to the Paine mechanism. Under mainland conditions where animals are free to disperse and are at risk of predation, densities of all rodent species might be held to low enough levels to reduce or eliminate interspecific aggression between them, thereby permitting coexistence (Grant 1972).

Thus there is limited evidence that the Paine effect may operate among certain terrestrial consumer guilds, but demonstrating it seems to require rather exacting conditions: predator-free environments and strong interspecific aggression within the guild of consumers. We therefore doubt that the Paine effect has much conservation significance in terrestrial communities except perhaps on predator-free islands where, in many cases, ecological conditions have already deteriorated beyond repair. As we shall see, the Paine effect may operate more commonly at the producer level of terrestrial and benthic ecosystems through changes in the abundance of consumers.

PREDATOR INTRODUCTION. Another kind of uncontrolled experiment is performed when predators are intentionally or unintentionally introduced into predator-free environments. The recovery of the sea otter from near extinction is a classic example. In the absence of sea otters, sea urchins, abalones, and other benthic grazers had nearly eliminated the kelp forests that once dominated the inshore environment along the Pacific rim of North America. Gradual recovery of the sea otter during the middle portion of the twentieth century has led to sharp declines of benthic grazers, accompanied by dramatic recovery of kelp forests and associated fauna (Estes et al. 1978, 1989). Experimental removal of benthic grazers, simulating otter predation, led to rapid growth of benthic algae, followed by progressive domination of a single kelp species, *Laminaria groenlandica,* demonstrating a strong Paine effect at the level of herbivore/plant interactions (Duggins 1980).

The introduction of alien top predators has wreaked havoc in freshwater aquatic systems around the world. Some particularly notorious cases are the introductions of sea lamprey to the Great Lakes, of Nile perch to Lake Victoria in East Africa, of rainbow trout to Lake Titicaca in the Andes, and of peacock bass to Lake Gatun, Panama. (See Zaret and Paine 1973; Zaret 1980; Kaufman 1992; Goldschmidt et al. 1993; Mills et al. 1994.) In these and countless additional well-documented examples, top-down effects have been dramatic and unequivocal—typically with devastating consequences for native fauna.

The introduction of exotic predators to predator-free islands provides further evidence for the operation of top-down regulation. Mongooses introduced onto islands of the tropical Pacific and Antilles have contributed to the collapse of native faunas (King 1984). Inadvertent introduction of the brown tree snake onto Guam led to a population explosion of the snake and consequent extinction of most of the island's native birds (Savidge 1987). Introduced domestic cats have had strong effects in Australia and on certain temperate islands, as have foxes in boreal to arctic regions (Bailey 1993).

On the North American mainland, the growing gray wolf population has been associated with a concurrent decline in elk and white-tailed deer densities. Most known ungulate mortality in these areas was caused by wolf predation (D. Pletscher, pers. com.). The recent reinvasion of the northern Midwest by wolves has reduced the distance from aquatic habitats that beavers can forage—a behavioral modification that in turn reduces the impact of beaver on plant associations (Naiman et al. 1994; Pollock et al. 1995). Similarly, the reestablishment of wolves in other areas has been followed by declines in caribou, moose, elk, and deer (Bergerud 1988; Messier and Crête 1985; Hatter and Janz 1994).


LONG-TERM MONITORING OF PREDATOR/PREY INTERACTIONS. A compelling case for a terrestrial trophic cascade is that of the gray wolf/moose/balsam fir interaction on Isle Royale, Michigan (McLaren and Peterson 1994; Messier 1994). The number of wolves determines the intensity of wolf predation on moose populations on Isle Royale. Growth rings in young fir trees showed depressed plant growth rates when wolves were rare and moose abundant—from which McLaren and Peterson (1994) infer the existence of a wolf-induced trophic cascade. Broad ramifications within the forest ecosystem are suggested from known linkages among moose, microbes, and soil nutrients (Pastor et al. 1988).

The anecdotal evidence cited here is consistent with top-down regulation as a predictable feature of terrestrial and many aquatic communities. But without rigorous controls, anecdotal evidence, by its nature, is open to alternative interpretations. Uncontrolled changes in the quality or distribution of habitats concurrent with predator elimination or reintroduction especially complicates the interpretation of causes and effects that may be separated in time by decades. For these reasons, scientists put greater stock in controlled comparisons and experiments.


### Experimental Evidence

Few well-controlled comparisons of prey populations at sites with and without top predators have been made—presumably because the conditions required are so rarely available. The sites being compared must have similar climate and vegetation and differ only in the presence/ absence of top predators. Hunting or complicating management interventions must be absent.

One carefully documented comparison is between two sites in the neotropics: one is Barro Colorado Island (BCI), Panama, a research preserve of the Smithsonian Institution; the other is Cocha Cashu Biological Station (CCBS) in the Manu National Park of Perú. Located respectively at 10° north and 12° south latitude, the two sites have a similar climate and fauna. The dominant habitat at both is primary tropical moist forest. BCI is a 1600-hectare island created by flooding during the construction of the Panama Canal. It has been isolated since the canal's creation. Due to its limited area, BCI lost top predators—jaguar, puma, and harpy eagle—more than fifty years ago (Glanz 1982). CCBS is located in the heart of a 2-million-hectare biosphere reserve that retains an intact flora and fauna, including all top predators.

The terrestrial and arboreal mammals of both BCI and CCBS have been censused on multiple occasions (Glanz 1990; Janson and Emmons 1990; Wright et al. 1994). Counts made by different observers at different times consistently agree in registering higher mammal densities on BCI than at CCBS (Terborgh 1988, 1992; Wright et al. 1994). In several cases, the differences in abundance are striking—exceeding an order of magnitude, particularly for the agouti, paca, armadillo, and coatimundi (terrestrial) and the three-toed sloth and tamandua (arboreal). Differences for other species are less extreme—as for the collared peccary and rabbit (terrestrial) and howler monkey (arboreal)—or negligible (deer, tapir). Whenever there are appreciable differences, they consistently favor BCI.

Differences in abundance are most pronounced in medium to large species that are prey of the top predators missing from BCI. Small mammals (rodents and marsupials weighing less than 1 kilogram) show similar abundances at the two sites, but these species do not appear in the prey of the top predators (Rettig 1978; Emmons 1987). Instead these animals are prey to small carnivores (ocelot, snakes, raptors) that are well represented at both sites. The higher densities of medium and large mammals on BCI have been interpreted as evidence of a top-down effect resulting from missing top predators (Terborgh and Winter 1980; Terborgh 1988, 1992). This conclusion, however, has been questioned by Wright et al. (1994) who emphasize that other interpretations are possible, including uncontrolled differences in productivity between the two sites.

The only certain way to exclude possible influences of uncontrolled variables is with strictly controlled experiments that include censusing before and after. For terrestrial predator/prey systems, the appropriate spatial scale on which to conduct the critical experiments is that of square kilometers—a fact that has precluded such experiments until very recently (Englund 1997). There are now two experimental efforts under way that promise to overcome certain weaknesses of correlational analysis and geographical comparisons. One of these efforts employs isolated remnants of a formerly intact landscape; the other uses large (1 square kilometer) fenced exclosures to exclude terrestrial predators. For reasons to be explained, neither set of experiments is perfect. But both represent major advances over previous efforts to isolate the effects of predators on terrestrial communities.

The creation in 1986 of one of the world's largest hydroelectric impoundments (Lago Guri) in the Caroni Valley of east-central Venezuela has resulted in the inundation of a hilly forested landscape with the consequent isolation of hundreds of erstwhile hilltops as islands. The impoundment is 120 kilometers long and up to 70 kilometers wide. Islands ranging in size from less than 1 hectare to more than 1000 hectares are scattered throughout the vast expanse of water—a number of them as far as 7 kilometers from the mainland. Small size and isolation by water assure that many of the more remote islands in Lago Guri are free of vertebrate predators except for certain small raptors and, perhaps, snakes.

Systematic surveys of the vertebrate faunas of a dozen Lago Guri islands were conducted seven years after isolation, along with control surveys on the nearby mainland (Terborgh et al. 1997). Roughly 75 to 90 percent of the species of terrestrial vertebrates that occupy the same forest type on the mainland were absent from islands between 1 and 10 hectares in size within seven years after isolation. With few exceptions, species that persisted became hyperabundant compared to their densities on the mainland. The absence of many species and the hyperabundance of others has created animal communities unlike any that would ever occur naturally—communities that are grotesquely imbalanced from a functional standpoint. These communities lack vertebrate predators and are deficient in pollinators and seed dispersers; but they contain abnormally high densities of seed predators (small rodents) and generalist herbivores (howler monkeys, iguanas, and leaf-cutter ants). The excess of herbivores is particularly striking, as all three species occur at densities between one and two orders of magnitude above those found on the mainland.

Larger Lago Guri islands (between 100 and 1000 hectares) still retain nearly complete vertebrate faunas (all primates and ungulates known for the region, for example), lacking only resident populations of the top predators (jaguar, puma, harpy eagle). Mammal densities on the two large islands being monitored have not yet increased conspicuously, but one and perhaps both of these islands are visited regularly by jaguars that swim over from the mainland, so they are not strictly predator-free. As for the smaller, more isolated islands that assuredly are predator-free, the hyperabundance of persistent vertebrates is consistent with the top-down effect of release from predation.

Further support for these observations is the documentation of hyperabundant rodent populations on numerous predator-free islands in both temperate and tropical regions (Adler and Levins 1994; Adler 1996). Nevertheless, the possibility remains of a confounding effect of missing species. The absence of other seed predators and herbivores that are present in the mainland fauna, for example, may have made available additional resources that allowed the persistent species to achieve hyper-abundance. As in the previous examples considered here, the findings are consistent with a top-down effect but an airtight case remains elusive.

Finally, we come to the most carefully constructed test of top-down regulation conducted to date. Charles Krebs, Tony Sinclair, and their associates are conducting the experiment in southern Yukon, Canada, where they have been monitoring snowshoe hare populations for nearly a decade in 1-square-kilometer plots. Two of the plots are surrounded by electric fencing that excludes mammalian predators but is permeable to hares. Plots have been assigned to five treatments: control, food supplementation, fertilizer, predator exclusion, and predator exclusion with food supplementation (Krebs et al. 1995). Hares exhibited strong positive demographic responses to food supplementation and (partial) predator exclusion while continuing to follow the classic ten-year cycle of abundance. Averaged over the peak and decline phases, hare density was double that of controls under predator exclusion, triple with food supplementation, and eleven times greater under predator exclusion coupled with food supplementation (Krebs et al. 1995). The results strongly implicate both bottom-up and top-down regulation. This interpretation is complicated, however, by the free passage of hares in and out of predator exclosures and by the exposure of hares within exclosures to predation by goshawks and great horned owls. Nevertheless, the effort represents a bold attempt to conduct an experimental test of bottom-up and top-down regulation on an appropriate spatial scale with a natural predator/prey system.

Another series of large-scale experiments has been conducted to test the role of top-down regulation in freshwater aquatic systems (Carpenter and Kitchell 1993). Entire lakes in Wisconsin have been seined free of piscivorous or planktivorous fishes and the respective hauls exchanged between lakes in a series of dramatic whole-lake perturbations (Carpenter et al. 1985; Carpenter and Kitchell 1988). Removal of piscivorous fish (large-mouthed bass, the top carnivore in this system) leads to order-of-magnitude increases in planktivorous fish, decreases in the size and number of zooplankton (cladocerans), and strong increases in the standing crop of phytoplankton in a textbook top-down trophic cascade.

A variety of efforts designed to assess the polarity of trophic regulation in terrestrial and aquatic ecosystems have consistently produced results consonant with strong top-down effects. To date, however, most or all of these efforts have fallen short of making an airtight case because of the overwhelming logistical challenge of removing or excluding only the guild of top predators without altering anything else. Carpenter's studies of Wisconsin lakes provide the most unambiguous evidence. On land, perhaps the closest approximation yet achieved to the ideal experimental condition is found in areas like BCI in Panama and in North American parks and suburbs where mammal communities complete except for top predators live under protection from hunting (McShea et al. 1997). In both these situations, densities of medium and large mammals are much higher than can be considered normal, though other potentially complicating factors preclude drawing an unequivocal link to missing predators.

Admittedly many questions remain to be answered by future research. Nevertheless, in the spirit of metaanalysis, if one considers the entire collection of controlled and uncontrolled comparisons and experiments cited here, the consonance of the results suggests a much stronger conclusion than a single case standing alone. With so much evidence pointing in the same direction, the conclusion that top predators play a major regulatory role seems inescapable.

## Countercurrents

Although the evidence that top predators commonly limit the densities of their prey is compelling, one would be wrong to conclude that predators limit the numbers of all consumers. There are a variety of situations in nature that allow consumers to escape predation to varying degrees—often to the extent that top-down control by large carnivores does not operate. These probable exceptions, as we shall see, include both megaherbivores and herd-forming migratory ungulates. Moreover, one should not assume that because top predators play major roles in regulating prey populations in many ecosystems, they play equivalent roles in all ecosystems.

Prior to the late-Pleistocene and Holocene megafaunal overkill, nearly every ecosystem on earth included very large herbivore species too big (at least as adults) to be killed by the largest carnivores in the system. The prime living example is that of elephants, which were once distributed on all continents (except Australia and Antarctica) and a number of islands. Nearly all the earth's once abundant megaherbivores have been driven to extinction and only a few survive (Martin and Klein 1984). In Africa there are rhinos and hippos, in addition to elephants, that, as adults, enjoy immunity to lions. In the north, adult moose repel gray wolves; in the neotropical forest, tapirs shrug off jaguars. Elsewhere, Madagascar had its elephant birds, New Zealand its moas, the Antilles their hutias and ground sloths, and the Seychelles, Galápagos and Aldabra

Island their tortoises. Lacking any population control from the top, megaherbivores must be regulated from below. But to the extent that megaherbivores regulate vegetation, they too exert a top-down force, that is independent of predation (Kortlandt 1984; Owen-Smith 1988). What fraction of the earth's land surface still supports megaherbivores? Ubiquitous and abundant to the point of dominating mammalian biomass over most of the globe for millions of years, megaherbivores have been so systematically persecuted that they have become almost irrelevant to today's ecosystems and conservation concerns, except in dwindling portions of Africa and Asia.

Sheer size enables a few of the world's largest mammals to escape predation. But size is not the only successful antipredator strategy to have arisen through evolution. Some species are able to reduce (but not eliminate) predation through social mechanisms. The list of these mechanisms is long. It includes the formation of herds and flocks, sentinel behavior, and the giving of alarm calls (Bertram 1978; Harvey and Greenwood 1978; Terborgh 1990). Social mechanisms can be very effective at limiting predation. Consider the fabled wildebeest of Serengeti. These antelopes aggregate in huge mixed herds that can be within the territories of only one or two lion prides at a time. Lions are consequently unable to make much of a dent in wildebeest numbers, killing only about 8 percent of the population per year (Sinclair and Norton-Griffiths 1979; Sinclair and Arcese 1995) In a bad year, wildebeest die en masse from starvation and malnutrition, as has been convincingly documented by Sinclair and his associates. The conclusion follows that wildebeest—and, by analogy, other herd-forming migratory ungulates—are regulated from the bottom up (Fryxell et al. 1988). But again, how much of today's earth is occupied by herd-forming migratory ungulates? Not much more than is occupied by megaherbivores. Both of these major agents of top-down forces in terrestrial ecosystems are becoming Pleistocene relics. Hence we should give special attention to top carnivore processes, because it seems likely that they are crucial to preserving what bits and pieces of wild nature we have left.

Top predators play structuring roles in many ecosystems. Exceptions, however, may be found in extreme environments, such as deserts or barrens, where low plant productivity or chemical toxicity of foliage limits large herbivores to such a degree that predators are unable to exploit them. Other factors, such as a severe disturbance, can temporarily upset normal trophic relationships. A stand-replacing fire, for example, may result in lowered herbivore densities and a switch from top-down to bottom-up regulation until the vegetation recovers (McLaren and Peterson

1994). In the world at large, however, productivity-limited (pure bottom-up) systems appear to be rare. Moderate to strong top-down regulation appears to be the norm for terrestrial ecosystems.

## Indirect Effects and Trophic Cascades

Having made a case for top-down regulation as a nearly ubiquitous force in terrestrial ecosystems, we now ask about the role played by top predators in maintaining ecosystem integrity. From a conservation perspective, we are concerned about the destabilizing forces that are unleashed in ecosystems from which top predators have been eliminated. It is a concern that extends over the large fraction of the earth's surface from which we have diminished or expunged the influence of these key animals. If there are no predictable ecological consequences of predator loss, we need not be concerned. But we have already reviewed convincing evidence to the contrary, so we know there are consequences. What are these consequences and how severe might they be?

The intellectual groundwork for studying "indirect effects" or "trophic cascades" in terrestrial ecosystems was laid in the 1970s and 1980s by James Brown and Diane Davidson in a major series of exclosure experiments conducted in the Chihuahuan Desert of southeastern Arizona. Experimental enclosures were open to aerial predators and certain mammals (coyotes) but closed to certain terrestrial predators (snakes) and to the movements of small rodents. Treatments included open and enclosed control plots, plus food supplementation and removal of rodents, ants, and both rodents and ants. (See Brown and Davidson 1977; Brown et al. 1986; Heske et al. 1994.) Rodents and ants live at the same trophic level: both subsist on the seeds of desert plants.

Partial exclusion of rodent predators led to increased densities of rodents, but not of ants. Selective removal of rodents or ants (or both) resulted in changes in the abundance and species composition of annual plants. In short, manipulation of a guild of consumers, in this case seed predators, resulted in large and often unanticipated changes in the composition of the plant community. Integrity of plant communities is essential to preserving biodiversity, so the Brown and Davidson experiments raised an early warning flag to conservationists. Perhaps other changes in consumer guilds mediated through top-down effects could have similarly drastic consequences.

In many parts of North America, extirpation of dominant predators has resulted in a phenomenon known as "mesopredator release" in areas supporting other small to midsized predators (foxes, skunks, raccoons, opossums, feral and domestic housecats: Soulé et al. 1988; Palomares et al. 1995). In such areas, mesopredators act by default as surrogate top predators. This has resulted in modified niche exploitation, altered diversity, and other ripple effects in the population structure of the community. Local elimination of coyotes, for example, allows the guild of mesopredators to increase in number, thereby imposing added predator pressure on the prey. Widespread reduction of ground-nesting birds, such as quail, pheasants, grouse, ducks, nightjars, and certain warblers, has been attributed to mesopredator release (Côté and Sutherland 1997). Mesopredator release has also been blamed for the decline or disappearance of gamebirds, songbirds, and other small vertebrates from a number of North American terrestrial ecosystems—including scrub habitats (Soulé et al. 1988), grasslands (Vickery et al. 1994), prairie wetlands (Sovada et al. 1995; Garrettson et al. 1996a, 1996b), and eastern deciduous forest (Wilcove 1985; Faaborg et al. 1995; Peterjohn et al. 1995).

Reintroduction or recolonization of predators influences the composition and structure of carnivore guilds as well. Wolf recovery in the Rocky Mountains has resulted in interference and exploitation competition among intraguild carnivores, resulting in changes in behavior, abundance, and distribution of affected species (Cohn 1998). As a rule, generalized predators, like the wolf, can be expected to exert stronger top down effects than specialists like the fisher and pine martin or omnivores such as bears.

Extirpation of top predators has released herbivore populations in parts of the United States with consequences that are just beginning to come to light. Overbrowsing by white-tailed deer is decisively altering the pattern of tree regeneration in some eastern forests and is threatening certain endangered plants with extinction (Alverson et al., 1988, 1994; Miller et al. 1992; McShea et al. 1997; Rooney and Dress 1997). Elsewhere in North America, introduced ungulates, especially Eurasian boar *(Sus scrofa),* have increased to such a degree that they are destroying wildflower beds and altering tree regeneration patterns in forests (Abramson 1992). It hardly needs to be emphasized that rapid, large-scale, and unpredictable changes in forest composition represent a chilling threat to biodiversity.

For another case, let us return to Lago Guri in Venezuela, where recently created islands in a hydroelectric impoundment are experiencing cataclysmic biological change. In a predator-free environment, three generalist herbivores have each increased in abundance by more than an order of magnitude. Howler monkeys on some islands have attained densities equivalent to 500 per square kilometer whereas mainland densities are typically between 20 and 40 per square kilometer (Crockett and Eisenberg 1986). Densities of iguanas and leaf-cutter ants have similarly exploded (Terborgh et al. 1997; Rao 1998).

Ongoing studies of forest regeneration on these islands reveal little successful reproduction of canopy trees. On some islands fewer than five species are represented by saplings in the understory, despite the presence of sixty to seventy species in the canopy. The mechanisms by which tree reproduction on these islands is being suppressed are currently under investigation. Preliminary results suggest the simultaneous involvement of several mechanisms: deficiencies of pollination and seed dispersal; excessive seed predation; decimation of seedlings by leaf-cutter ants; and repeated defoliation of canopy trees by howler monkeys, iguanas, and leaf-cutter ants (Terborgh et al., unpublished results). In the absence of "normal" biological interactions, the remnant ecosystems of these islands have spun out of control. It seems inevitable that most of the plant and animal species that survived the initial contraction in area will go extinct within one or two tree replacement cycles.

Vegetation change in the Lago Guri islands and in portions of the United States occupied by hyperabundant populations of white-tailed deer and Eurasian boar offer startling examples of trophic cascades—examples that mirror findings from deserts (Brown et al. 1986), lakes (Carpenter and Kitchell 1993), and Pacific kelp forests (Estes et al. 1989). To prevent ecosystems all over North America from experiencing similar convulsions brought about by trophic cascades, the full spectrum of ecological processes that operates to perpetuate biodiversity, especially predation, must be widely maintained.

Where top predators have been extirpated and their reestablishment is impractical, can trophic cascades be avoided? Perhaps worst-case scenarios can be avoided through interventions of various sorts. But no human effort can accurately simulate the effects of real predators, because these animals have impacts on many prey species simultaneously and interact with prey populations in complex ways that are seldom understood. Nevertheless, the worst consequences of trophic cascades might be forestalled or ameliorated though the hunting of herbivores and trapping of mesopredators. The most severe impacts of hyperabundant mesopredators and consumers appear in localities where predators are absent and hunting and trapping are prohibited.

A contrasting situation arises in countries lacking enforced game laws, where all medium and large birds and mammals are systematically overhunted (Redford 1992). The resulting "defaunation," like hyperabundance, results in distorted or disrupted plant/animal interactions—including seed dispersal, seed predation, and herbivory. Little is known about the consequences of wholesale defaunation, though preliminary evidence from Mexico points to highly aberrant patterns of plant regeneration (Dirzo and Miranda 1991).

Predators prevent prey populations and mesopredators from exploding into hyperabundance while rarely, if ever, driving prey to extinction. Prey species, such as seed dispersers, seed predators, or herbivores, are thereby regulated within definite upper and lower bounds. The operation of such feedback mechanisms can be likened to "a balance of nature." Nature stays in balance so long as a fauna remains intact and the full suite of ecological processes operates unhindered. It is when nature falls out of balance—when there are too many consumers and mesopredators (or not enough)—that species begin to disappear and humans begin to notice. But what humans notice is only that some favored species or another has disappeared. Hidden in the workings of a nature we are only beginning to understand, the cause remains obscure.

## Another Key to Biodiversity

Despite the complexity of food web linkages, interactions across trophic levels define a subset of these links that are of particular importance to the functioning of natural ecosystems. In terrestrial ecosystems, top-down and bottom-up processes operate simultaneously. This seemingly contradictory statement results not only from the complexity of food web structure but from flexibility in the behavior of individual species—such as the tendency for prey to act as time-minimizers in the presence of predators and the ability of plants to increase their investment in anti-herbivore defenses in response to herbivory.

Although megaherbivores (those large enough to be invulnerable to predators) and herd-forming migratory ungulates tend to be regulated from the bottom up, megaherbivores concurrently exert top-down forces through their effects on vegetation. Both groups of species may have been prominent over much of the earth's surface prior to megafaunal overkill, but they have been reduced by human persecution to a tiny fraction of their former geographical occurrence. What remains nearly everywhere else are drastically truncated mammal communities that are regulated largely through top-down processes.

The evidence reviewed here overwhelmingly supports the strong top-down role of top carnivores in regulating prey populations—and thereby stabilizing the trophic structure of terrestrial ecosystems. Loss of top predators results in hyperabundance of consumers playing a variety of trophic roles (herbivores, seed dispersers, seed predators) and in mesopredator release. Hyperabundance of consumers and mesopredators, in turn, results in trophic cascades that lead to multiple effects—including the direct elimination of plant populations from overbrowsing/grazing, reproductive failure of canopy tree species, and the loss of ground-nesting birds and probably other small vertebrates.

In sum, then, our current knowledge about the natural processes that maintain biodiversity suggests a crucial and irreplaceable regulatory role of top predators. The absence of top predators appears to lead inexorably to ecosystem simplification accompanied by a rush of extinctions. Therefore, efforts to conserve North American biodiversity in interconnected mega-reserves will have to place a high priority on reestablishing top predators wherever they have been locally extirpated. If steps are not taken in the interim to restore the full gamut of natural abiotic and biotic processes that maintain biodiversity, efforts to halt extinction through legislated mechanisms (such as the Endangered Species Act) will be overwhelmed by irresistible biological forces. It is only by providing the conditions that allow nature to remain in balance that biodiversity can be perpetuated over the long run.

# References

Abrams, P. A. 1993. Why predation rate should not be proportional to predator density. *Ecology* 74:726–733.

Abramson, R. 1992. Rooting out wild pigs' mischief in Great Smoky Mountains. *Los Angeles Times,* 21 July 1992, p. A5.

Adler, G. H. 1996. The island syndrome in isolated populations of a tropical forest rodent. *Oecologia* 108:694–700.

Adler, G. H., and R. Levins. 1994. The island syndrome in rodent populations. *Quarterly Review of Biology* 69:473–490.

Adler, G. H., and J. O. Seamon. 1991. Distribution and abundance of a tropical rodent, the spiny rat, on islands in Panama. *Journal of Tropical Ecology* 7:349–360.

Alverson, W. S., W. Kuhlmann, and D. M. Waller. 1994. *Wild forests: Conservation biology and public policy.* Washington, D.C.: Island Press.

Alverson, W. S., D. M. Waller, and S. L. Solheim. 1988. Forests too deer: Edge effects in northern Wisconsin. *Conservation Biology* 2:348–358.

Bailey, E. P. 1993. Introduction of foxes to Alaskan islands—history, effects on avifauna, and revocation. Resource publication 193. Washington, D.C.: U.S. Department of the Interior, Fish and Wildlife Service.

Bergerud, A. T. 1988. Caribou, wolves, and man. *Trends in Ecology and Evolution* 3:68–72.

Berry, R. J., and B. J. K. Tricker. 1969. Competition and extinction: The mice of Foula, with notes on those of Fair Isle and St. Kilda. *Journal of Zoology* (London) 158:247–265.

Bertram, B. C. R. 1978. Living in groups: Predators and prey. In J. R. Krebs and N. B. Davies (eds.), *Behavioral ecology.* Sunderland, Mass.: Sinauer.

Bowen, L., and D. van Vuren. 1997. Insular endemic plants lack defenses against herbivores. *Conservation Biology* 11:1249–1254.

Bramwell, D. (ed.). 1979. *Plants and islands.* London: Academic Press.

Brett, M. T., and C. R. Goldman. 1997. Consumer versus resource control in freshwater pelagic food webs. *Science* 275:384–386.

Brosset, A. 1963. Statut actuel des mammifères des îles Galápagos. *Mammalia* 27:323–338.

Brown, J. H., and D. W. Davidson. 1977. Competition between seed-eating rodents and ants in desert ecosystems. *Science* 196:880–882.

Brown, J. H., D. W. Davidson, J. C. Munger and R. S. Inouye. 1986. Experimental community ecology: The desert granivore system. In J. Diamond and T. J. Case (eds.), *Community ecology.* New York: Harper and Row.

Carlquist, S. 1974. *Island biology.* New York: Columbia University Press.

Carpenter, S. R., and J. F. Kitchell, 1988. Consumer control of lake productivity. *BioScience* 38:764–769.

———. 1993. *The trophic cascade in lakes.* Cambridge: Cambridge University Press.

Carpenter, S. R., J. F. Kitchell and J. R. Hodgson. 1985. Cascading trophic interactions and lake productivity. *BioScience* 35:634–639.

Caughley, G. 1994. Directions in conservation biology. *Journal of Animal Ecology* 63:215–244.

Coblentz, B. E. 1978. Effects of feral goats *(Capra hircus)* on island ecosystems. *Biological Conservation* 13:279–286.

———. 1980. Effects of feral goats on the Santa Catalina Island ecosystem. In D. M. Power (ed.), *California islands: Proceedings of a multidisciplinary symposium.* Santa Barbara: Museum of Natural History.

———. 1990. Exotic organisms: A dilemma for conservation biology. *Conservation Biology* 4:261–265.

Cohn, J. 1998. A dog-eat-dog world? *BioScience* 48:430–434.

Connell, J. H. 1978. Diversity in tropical rain forests and coral reefs. *Science* 199:1302–1310.

Coope, G. R. 1995. Insect faunas in ice age environments: Why so little extinction? In J. H. Lawton and R. M. May (eds.), *Extinction rates.* Oxford: Oxford University Press.

Côté, I. M., and W. J. Sutherland. 1997. The effectiveness of removing predators to protect bird populations. *Conservation Biology* 11:395–405.

Crockett, C. M., and J. F. Eisenberg. 1986. Howlers: Variations in group size and demography. In B. B. Smuts, D. L. Cheney, R. M. Seyfarth, R. W. Wrangham, and T. T. Struhsaker (eds.), *Primate societies.* Chicago: University of Chicago Press.

Cronk, Q. C. B. 1980. Extinction and survival in the endemic vascular flora of Ascension Island. *Biological Conservation* 13:207–219.

Crosby, A. W. 1986. *Ecological imperialism.* Cambridge: Cambridge University Press.

Diamond, J. M. 1982. Man the exterminator. *Nature* 298:787–789.

Dirzo, R., and A. Miranda. 1991. Altered patterns of herbivory and diversity in the forest understory: A case study of the possible consequences of contemporary defaunation. In P. W. Price, P. W. Lewinsohn, G. W. Fernandes, and W. W. Benson (eds.), *Plant-animal interactions: Evolutionary ecology in tropical and temperate regions.* New York: Wiley.

Duggins, D. O. 1980. Kelp beds and sea otters: An experimental approach. *Ecology* 61:447–453.

Ehrlich, P. R. 1995. The scale of the human enterprise and biodiversity loss. In J. H. Lawton and R. M. May (eds.), *Extinction rates.* Oxford: Oxford University Press.

Emmons, L. H. 1987. Comparative feeding ecology of felids in a neotropical rainforest. *Behavioral Ecology and Sociobiology* 20:271–283.

Englund, G. 1997. Importance of spatial scale and prey movements in predator caging experiments. *Ecology* 78(8):2316–2325.

Erlinge, S., G. Göransson, G. Högstedt, G. Jansson, O. Liberg, J. Loman, I. N. Nilsson, T. von Schantz, and M. Sylvén. 1984. Can vertebrate predators regulate their prey? *American Naturalist* 123:125–133.

———. 1988. More thoughts on vertebrate predator regulation of prey. *The American Naturalist* 132:148–154.

Estes, J. A. 1996. Predators and ecosystem management. *Wildlife Society Bulletin* 24:390–396.

Estes, J. A., D. O. Duggins, and G. B. Rathbun. 1989. The ecology of extinctions in kelp forest communities. *Conservation Biology* 3:252–264.

Estes, J. A., N. S. Smith, and J. F. Palmisano. 1978. Sea otter predation and community organization in the western Aleutian Islands, Alaska. *Ecology* 59:822–833.

Faaborg, J., M. Brittingham, T. Donovan, and J. Blake. 1995. Habitat fragmentation in the temperate zone. In T. E. Martin and D. M. Finch (eds.), *Ecology and management of neotropical migratory birds: A synthesis and review of critical issues.* New York: Oxford University Press.

Fryxell, J. M., J. Greever, and A. R. E. Sinclair. 1988. Why are migratory ungulates so abundant? *American Naturalist* 131:781–798.

Garrettson, P. R., F. C. Rohwer, L. A. Jones, and B. J. Mense. 1996a. Predator management to benefit prairie-nesting ducks. In J. T. Ratti (ed.), *Seventh International Waterfowl Symposium.* Memphis: Ducks Unlimited.

Garrettson, P. R., F. C. Rohwer, J. M. Zimmer, B. J. Mense, and N. Dion. 1996b. Effects of mammalian predator removal on waterfowl and non-game birds in North Dakota. In *Transactions of the 61st North American Wildlife and Natural Resources Conference.*

Garrott, R. A., P. J. White, and C. A. Vanderbilt White. 1993. Overabundance: An issue for conservation biologists? *Conservation Biology* 7:946–949.

Glanz, W. E. 1982. Adaptive zones of neotropical mammals: A comparison of some temperate and tropical patterns. In M. A. Mares and H. H. Genoways (eds.), *Mammalian biology in South America*. Special publication of Pymatuning Laboratory of Ecology. Pittsburgh: University of Pittsburgh.

————. 1990. Neotropical mammal densities: How unusual is the community of Barro Colorado Island, Panama? In A. H. Gentry (ed.), *Four neotropical rainforests*. New Haven: Yale University Press.

Goldschmidt, T., F. Witte, and J. Wanink. 1993. Cascading effects of the introduced Nile perch on the detritivorous/phytoplanktivorous species in the sublittoral areas of Lake Victoria. *Conservation Biology* 7(3):686–700.

Goodman, S. M. 1995. *Rattus* on Madagascar and the dilemma of protecting the endemic rodent fauna. *Conservation Biology* 9:450–453.

Grant, P. R. 1972. Interspecific competition among rodents. *Annual Review of Ecology and Systematics* 3:79–106.

Hairston, N. G., F. E. Smith, and L. B. Slobodkin. 1960. Community structure, population control, and competition. *American Naturalist* 94:421–424.

Harvey, P. H., and P. J. Greenwood. 1978. Anti-predator defence strategies: Some evolutionary problems. In J. R. Krebs and N. B. Davies (eds.), *Behavioural ecology*. Sunderland, Mass.: Sinauer.

Hatter, I., and D. W. Janz. 1994. The apparent demographic changes in black-tailed deer associated with wolf control in northern Vancouver Island, Canada. *Canadian Journal of Zoology* 72:878–884.

Heske, E. J., J. H. Brown, and S. Mistry. 1994. Long-term experimental study of a Chihuahuan desert rodent community: 13 years of competition. *Ecology* 75(2):438–445.

Hunter, M. D., and P. W. Price. 1992. Playing chutes and ladders: Heterogeneity and the relative roles of bottom-up and top-down forces in natural communities. *Ecology* 73(3):724–732.

Janson, C. H., and L. H. Emmons. 1990. Ecological structure of the nonflying mammal community at Cocha Cashu Biological Station, Manu National Park, Peru. In A. H. Gentry (ed.), *Four neotropical rainforests*. New Haven: Yale University Press.

Kaufman, L. 1992. Catastrophic change in a species-rich freshwater ecosystem: Lessons from Lake Victoria. *BioScience* 42:846–858.

Kidd, N. Z. C., and G. B. Lewis. 1987. Can vertebrate predators regulate their prey? A reply. *American Naturalist* 130:448–453.

King, C. M. 1984. *Immigrant killers: Introduced predators and the conservation of birds in New Zealand*. Aukland: Oxford University Press.

Klein, D. R. 1968. The introduction, increase, and crash of reindeer on St. Matthew's Island. *Journal of Wildlife Management* 32:350–367.

Kortlandt, A. 1984. Vegetation research and the "bulldozer" herbivores of tropical Africa. In A. C. Chadwick and C. L. Sutton (eds.), *Tropical rainforest*. Leeds: Leeds Philosophical Literature Society.

Krebs, C. J., S. Boutin, R. Boonstra, A. R. E. Sinclair, J. N. M. Smith, M. R. T. Dale, K. Martin, and R. Turkington. 1995. Impact of food and predation on the snowshoe hare cycle. *Science* 269:1112–1118.

Lynam, A. J. 1997. Rapid decline of small mammal diversity in monsoon ever-green forest fragments in Thailand. In W. F. Laurance and R. O. Bierregaard Jr. (eds.), *Tropical forest remnants: Ecology, management, and conservation of frag-mented communities.* Chicago: University of Chicago Press.

Martin, P. S., and R. G. Klein (eds.). 1984. *Quaternary extinctions: A prehistoric rev-olution.* Tuscon: University of Arizona Press.

Matson, P. A., and M. D. Hunter. 1992. The relative contributions of top-down and bottom-up forces in population and community ecology. *Ecology* 73:723.

May, R. M., J. H. Lawton, and N. E. Stork. 1995. Assessing extinction rates. In J. H. Lawton and R. M. May (eds.), *Extinction rates.* Oxford: Oxford University Press.

McLaren, B. E., and R. O. Peterson. 1994. Wolves, moose, and tree rings on Isle Royale. *Science* 266:1555–1558.

McShea, W. J., H. B. Underwood, and J. H. Rappole. 1997. *The science of over-abundance: Deer ecology and population management.* Washington, D.C.: Smith-sonian Institution Press.

Menge, B. A. 1992. Community regulation: Under what conditions are bottom-up factors important on rocky shores? *Ecology* 73:755–765.

———. 1995. Indirect effects in marine rocky intertidal interaction webs: Pat-terns and importance. *Ecological Monographs* 65:21–74.

Menge, B. A., E. L. Berlow, C. A. Blanchette, S. A. Navarrete, and S. B. Yamada. 1994. The keystone species concept: Variation in interaction strength in a rocky intertidal habitat. *Ecological Monographs* 64:249–286.

Messier, F. 1994. Ungulate population models with predation: A case study with the North American moose. *Ecology* 75(2):478–488.

Messier, F., and M. Crête. 1985. Moose-wolf dynamics and the natural regulation of moose populations. *Oecologia* 65:503–512.

Miller, S. G., S. P. Bratton, and J. Hadidian. 1992. Impacts of white-tailed deer on endangered and threatened vascular plants. *Natural Areas Journal* 12:67–74.

Mills, E. L., J. H. Leach, J. T. Carlton, and C. L. Secor. 1994. Exotic species and the integrity of the Great Lakes. *BioScience* 44:666–676.

Naiman, R. J., G. Pinay, C. A. Johnston, and J Pastor. 1994. Beaver influences on the long-term biogeochemical characteristics of boreal forest drainage net-works. *Ecology* 75:905–921.

Oksanen, L. 1983. Trophic exploitation and arctic phytomass patterns. *American Naturalist* 122:45–52.

Owen-Smith, N. 1988. *Megaherbivores: The influence of very large body size on ecol-ogy.* Cambridge: Cambridge University Press.

Paine, R. 1966. Food web complexity and species diversity. *American Naturalist* 100:65–75.

Palomares, F., P. Gaona, P. Ferreras, and M. Debiles. 1995. Positive effects on game species of top predators by controlling smaller predator populations: An example with lynx, mongooses, and rabbits. *Conservation Biology* 9(2): 295–305.

Pastor, J., R. J. Naiman, and B. Dewey. 1988. Moose, microbes, and boreal forests. *BioScience* 38:770–777.

Peterjohn, B. G., J. R. Sauer, and C. S. Robbins. 1995. Population trends from the North American Breeding Bird Survey. In T. E. Martin and D. M. Finch (eds), *Ecology and management of neotropical migratory birds: A synthesis and review of critical issues.* New York: Oxford University Press.

Pielou, E. C. 1991. *After the ice age: The return of life to glaciated North America.* Chicago: University of Chicago Press.

Polis, G. A., and D. R. Strong. 1996. Food web complexity and community dynamics. *American Naturalist* 147:813–846.

Pollock, M. M., R. J. Naiman, H. E. Erickson, C. A. Johnston, J. Pastor, and C. Pinay. 1995. Beaver as engineers: Influences on biotic and abiotic characteristics of drainage basins. In C. G. Jones and J. H. Lawton (eds.), *Linking species and ecosystems.* New York: Chapman & Hall.

Power, M. E. 1992. Top-down and bottom-up forces in food webs: Do plants have primacy? *Ecology* 73:733–746.

Rao, M. 1998. Relaxation in the Lago Guri islands, Venezuela: Causes underlying leaf-cutter ant (*Atta* sp.) hyperabundance and consequences for plant species diversity. PhD dissertation, Duke University, Durham, N.C.

Redford, K. H. 1992. The empty forest. *BioScience* 42:412–422.

Rettig, N. L. 1978. Breeding behavior of the harpy eagle *(Harpia harpyja). Auk* 95:629–643.

Rooney, T. P., and W. J. Dress. 1997. Species loss over sixty-six years in the ground-layer vegetation of Heart's Content and old-growth forest in Pennsylvania, USA. *Natural Areas Journal* 17:297–305.

Savidge, J. A. 1987. Extinction of an island avifauna by an introduced snake. *Ecology* 68:660–668.

Sinclair, A. R. E., and P. Arcese. 1995. Population consequences of predation-sensitive foraging: The Serengeti wildebeest. *Ecology* 76:882–891.

Sinclair, A. R. E., and M. Norton-Griffiths (eds.). 1979. *Serengeti: Dynamics of an ecosystem.* Chicago: University of Chicago.

Soulé, M. E., E. T. Bolger, A. C. Alberts, J. Wright, M. Sorice, and S. Hill. 1988. Reconstructed dynamics of rapid extinctions of chaparral-requiring birds in urban habitat islands. *Conservation Biology* 2:75–92.

Sousa, W. P. 1984. Intertidal mosaics: Patch size, propagule availability, and spatially variable patterns of succession. *Ecology* 65:1918–1935.

Sovada, M. A., A. B. Sargeant, and J. W. Grier. 1995. Differential effects of coyotes and red foxes on duck nest success. *Journal of Wildlife Management* 59:1–9.

Stanley, S. M. 1987. *Extinction.* New York: Freeman Scientific American Library.

Strong, D. R. 1992. Are trophic cascades all wet? Differentiation and donor-control in speciose ecosystems. *Ecology* 73(3):747–754.

Terborgh, J. 1988. The big things that run the world—a sequel to E. O. Wilson. *Conservation Biology* 2:402–403.

———. 1990. Mixed flocks and polyspecific associations: Costs and benefits of mixed groups of birds and monkeys. *American Journal of Primatology* 21:87–100.

———. 1992. Maintenance of diversity in tropical forests. *Biotropica* 24:283–292.

Terborgh, J., and B. Winter. 1980. Some causes of extinction. In M. E. Soulé and B. A. Wilcox (eds.), *Conservation biology: An evolutionary-ecological approach.* Sunderland, Mass.: Sinauer.

Terborgh, J., L. Lopez, J. Tello, D. Yu, and A. R. Bruni. 1997. Transitory states in relaxing land bridge islands. In W. F. Laurance and R. O. Bierregaard Jr. (eds.), *Tropical forest remnants: Ecology, management, and conservation of fragmented communities.* Chicago: University of Chicago Press.

Vickery, P. D., M. L. Hunter Jr., and S. M. Melvin. 1994. Effects of habitat area on the distribution of grassland birds in Maine. *Conservation Biology* 8:1087–1097.

Vitousek, P. 1988. Diversity and biological invasions of oceanic islands. In E. O. Wilson and F. M. Peter (eds.), *Biodiversity.* Washington, D.C.: National Academy Press.

Wedin, D., and D. Tilman. 1993. Competition among grasses along a nitrogen gradient: Initial conditions and mechanisms of competition. *Ecological Monographs* 63:199–229.

Werner, E. E., and B. R. Anholt. 1993. Ecological consequences of the trade-off between growth and mortality rates mediated by foraging activity. *American Naturalist* 142:242–272.

Werner, E. E., and D. J. Hall. 1988. Ontogenetic habitat shifts in the bluegill sunfish *(Lepomis macrochirus):* The foraging rate-predation risk trade-off. *Ecology* 69:1352–1366.

Wilcove, D. 1985. Nest predation in forest tracts and the decline of migratory songbirds. *Ecology* 66:1211–1214.

Wilson, E. O. 1992. *The diversity of life.* Cambridge, Mass.: Harvard University Press.

Wright, S. J., M. E. Gompper, and B. de Leon. 1994. Are large predators keystone species in neotropical forests? The evidence from Barro Colorado Island. *Oikos* 71:279–294.

Zaret, T. M. 1980. *Predation and freshwater fish communities.* New Haven: Yale University Press.

Zaret, T. M., and R. T. Paine. 1973. Species introduction in a tropical lake. *Science* 182:449–455.

# 4 Regional and Continental Restoration

*Daniel J. Simberloff, Dan Doak, Martha Groom,
Steve Trombulak, Andy Dobson, Steve Gatewood,
Michael E. Soulé, Michael Gilpin, Carlos Martínez
del Rio, and Lisa Mills*

Goals of ecological restoration have been numerous. For academic ecologists, "restoration" is generally the exact reproduction, with human intervention, of a community or ecosystem that once was present (Magnuson et al. 1980; Bradshaw 1987). But even this goal entails a decision that is fundamentally normative rather than scientific. What previous community or ecosystem do we wish to reproduce: the one that existed a century ago, the one that existed before European colonization of North America, or the one that existed before human colonization of North America? Many activities that pass in the lay literature as restoration fall far short of the academic goal (Bradshaw 1987; Simberloff 1990; Cairns 1993). Rehabilitation, for example, is the partial reestablishment of some previous ecosystem, but with modifications. Revegetation and reclamation simply mean establishing some form of vegetation (as for erosion control). Recovery is partial restoration by secondary succession (without human assistance). Often activities termed "restoration" seem to be carried out without very specific goals, much less the exact reproduction of some previous system. The $55 million Batiquitos Lagoon Enhancement project, in a 240-hectare wetland on the southern border of Carlsbad, California, for example, is widely referred to by the press, public officials, and others as "restoration" (Rodgers 1997). Yet its goals seem unclear other than to remove accumulated sand and silt. Exactly which species are targeted for return, and why?

# Goals of Restoration

Much restoration ecology consists of empirical tools developed to deal with specific local problems, like erosion or pollution, largely because funding is available for such projects. Although important lessons for restoration, strictly defined, arise from this pragmatic, small-scale work, it is generally aimed at solving a specific, often nonecological, problem—such as a threat to human health or an aesthetic horror—rather than restoring a former community or ecosystem. Further, strong experimental design has not characterized this approach, and hypothesis testing in restoration is rare. Thus the field of restoration ecology has not evolved rapidly—despite widespread recognition (Jordan et al. 1987; Dobson et al. 1997) that restoration is vitally needed to deal with environmental and conservation problems and can be a powerful tool in improving our general ecological understanding.

A goal of many conservationists—and an aim of The Wildlands Project (TWP)—is to maintain the full range of native species and ecosystems. This is a much more ambitious goal than that of most restoration projects, and it is a pity that the field of restoration ecology cannot yet provide a set of models and tools to ensure its attainment. Restoration ecologists can, however, greatly aid conservation initiatives like TWP, and they, in turn, can learn much from restorations attempted under its aegis.

Whatever the restoration project, one must first choose a specific goal: the full range of native species and ecosystems *when*? This "reference time" is the first decision. Second, given a reference time, one must determine what is acceptably close to the full range of native species and ecosystems present at that time. In some instances, native species are extinct; in others, the logistical problems of rebuilding a species population or a community are insurmountable. In fact, one can never achieve an exact reproduction of the system that once obtained—if only because the individuals that existed at the reference time, with their unique genotypes, no longer exist (Simberloff 1990). Further, even pristine systems, undisturbed by humans, are dynamic. (The process of succession, for example, entails community change.) Thus restoration should be aimed at returning to the point on this trajectory of change that would have obtained in the absence of human disturbance, rather than trying to replicate the precise system that once was present (Atkinson 1990; Simberloff 1990). Of course, specifying what that trajectory would have been is an enormous scientific challenge. But in principle this should be our goal, and we must be able to state how closely our goal must be approximated in order for a restoration project to be called a success.

Third, given an overall restoration goal, criteria for success must be spelled out explicitly at both regional and local scales. These measures of success or progress will vary greatly. For TWP, what constitutes "wildness" in parts of the West is unattainable in many parts of the East if only because the human population density is too great. Nevertheless, even a fairly small and isolated area embedded in a developed landscape can, with proper management, be made to approximate a previous state more closely. The expense of such an endeavor may militate against making it a priority, as there will surely be economies of scale when many localities in a region can all be restored (Cairns 1993). If a project is undertaken, however, we must have clear criteria for what variables to monitor and how to use these data to judge whether our goal has or has not been achieved—or at least whether we are headed in the right direction. We also must consider the temporal component of each specific restoration project. How long do we expect a project to take? If the trend is in the right direction, is it occurring quickly enough to justify maintaining a course of action? In some situations, unaided secondary succession may move an ecosystem from a degraded state a substantial way toward its original condition. This, however, may take a very long time (Bradshaw 1987). Under what circumstances do we declare that progress is insufficient and a modified plan is needed?

Fourth, there is increasing recognition that the restoration of community structure (species composition and relative abundances) often depends on restoring various ecological processes and functions. Later we shall discuss this requirement in detail. Occasionally, however, the emphasis on process and function leads to their becoming the goal of management and restoration (Armstrong 1993; Simberloff in press)—that is, the processes and functions themselves become the valued elements, rather than the species. To a degree, the fact that species can substitute for one another in performing certain functions (Atkinson 1988; Walker 1991) is both useful—for example, when a former species is now extinct—and part and parcel of the dynamic nature of normal ecological succession. And certainly processes and functions that support target species must themselves become restoration targets. But process and function are no substitute for species; energy flows, materials cycle, and other processes obtain even in greatly degraded communities (as in rice paddies and corn fields). In fact, some processes, such as primary productivity, may be greater in species-poor communities than in the species-rich communities they replaced. Thus we must not let process become the goal of restoration efforts at the expense of species (Soulé 1994).

Finally, both regional and local goals must be capable of being made operational. It does little good to state that we wish to reestablish the full range of native species and ecosystems if we cannot specify the activities that will lead toward this goal. As we shall see, these activities may vary regionally and even within regions.

## The Question of Scale

Efforts to restore plant communities have focused largely on small spatial scales and have aimed at achieving local effects. The restoration or reclamation of mine tailings, for example, generally involves at most the reestablishment of plant communities over a few hectares for aesthetic purposes and prevention of erosion and transport of heavy metals (Daniels and Zipper 1988). Other restorations that focus almost exclusively on small areas include mines (Covert 1990; Ward et al. 1990; Chapman and Younger 1995), landfills (Handel et al. 1997), contaminated urban areas (Smith 1988), wetlands (Forbes 1993; Reinartz and Warne 1993; Middleton 1995), tidal marshes (Broome et al. 1988), salt marshes (Zedler 1988), sand dunes (Pickart 1990; Seliskar 1995), and riparian areas (Meda 1990).

At somewhat larger spatial scales are efforts that seek to restore examples of specific community types. For instance, 14,850 hectares of logged redwood (*Sequoia sempervirens*) forest were added to Redwood National Park (California); since that time, an active restoration project has been undertaken with the ultimate goal of returning all land to redwood forest (Belous 1984). The Lincoln Boyhood National Memorial in Indiana contains 20 hectares of hardwood forest now actively managed to return it to old-growth conditions (Hellmers 1983). At the University of Wisconsin Arboretum, the oldest restoration project in the United States entails the restoration of 200 hectares, primarily to prairie, but including conifer stands (Jordan 1983). The Caño Island Biological Reserve, located 15 kilometers off the southwestern coast of Costa Rica, includes more than 6 kilometers of shoreline that contain species-rich coral communities. Since 1984, several areas within the reef habitat have been under restoration by transplantation of live coral fragments from other parts of the reef (Guzmán 1991). Restoration of the Kissimmee River floodplain in Florida targets some 6000 hectares of broadleaf marsh and 2000 hectares of wet prairie wetlands (Toth 1993). This scale (5000 to 10,000 hectares) also typifies ambitious attempts to restore the entire biotic community to small islands off Australia, Bermuda, Mauritius, and New Zealand (Atkinson 1988; Towns et al. 1997).

Animal reintroductions, often a part of restoration, have usually taken place at small spatial scales, as well, and usually focus on what is needed to support minimum viable populations of the target species.

Planned reintroductions of animals must confront several questions (Gilpin 1987): What requirements must be met to provide sufficient genetic variation, proper sex ratios, and whatever social cohesion is necessary to ensure population viability? When in the restoration process is it best to reintroduce animal species, and does the optimal time depend on the location of the species in the food web? What aspects of the landscape and season of the year must be considered? What margin of safety is advisable to maximize the probability of success in a reintroduction while maintaining costs within reasonable limits?

In landscape-level or regional projects, such as those undertaken by TWP, the restoration focus must be on large spatial scales, in some cases hundreds of square kilometers, to maintain the full range of native species and ecosystems. Restoration of natural conditions over such large areas is important for several reasons. First, some species require large areas for survival. Individuals of some species may range widely. At an extreme, wolves (*Canis lupus*) have been known to move hundreds of kilometers (Fritts 1983), although pack home ranges are normally from a few tens to a few thousands of square kilometers (Mech 1970). Many carnivores are asocial and therefore are found at very low densities—for example, there are fewer than a thousand grizzly bears (*Ursus arctos*) in Montana, Wyoming, and Idaho (Primm 1996). Other species occur naturally in low population densities and therefore must be present over a large area to have population sizes large enough to be viable. Estimated densities of wolverines (*Gulo gulo*) in the northern Rockies, for instance, range from 0.5 to 1.5 per 100 square kilometers (Weaver et al. 1996).

Second, many processes are ecologically effective only over large areas. Disturbance agents in some landscapes, such as fire in the Intermountain West, often spread over several hundred square kilometers. The recent Yellowstone fire covered 395,570 hectares, including some unburned patches within the burn matrix (Balling et al. 1992). Natural hydrological patterns, including those of flooding, can often be restored only by focusing on entire watersheds. The restoration of some evolutionary processes, such as natural selection, requires populations in the thousands. And speciation in large animals requires persistence of viable, isolated populations for long periods (Soulé 1980).

Third, the dynamic, nondeterministic character of natural communities requires restoration of large areas in order to promote the long-term viability and adaptability of populations and communities. This character is to some extent a function of the large spatial scales over which some disturbance agents generate spatially heterogeneous patterns. The transition of the northeastern United States from late to early successional condition, for example, with respect to both species composition (Peterson and Pickett 1995) and physical structure (Foster et al. 1997), is strongly driven by rare but catastrophic storms. The exact locations of where storms will reset the successional clock are, however, unpredictable. If a restoration goal is to restore old age classes of forests, then restoration must be regional in order to minimize the chance that all of this forest type is lost in a single disturbance event (Trombulak 1996). The danger of highly local restoration is exemplified by the attempt to protect Cathedral Pines, an 11-hectare stand of old-aged white pines (*Pinus strobus*) in Connecticut. In 1989, a tornado destroyed the entire stand—effectively eliminating that community type from the entire region (Davis 1993). The effort to restore the Puerto Rican parrot (*Amazona vittata*) population was greatly jeopardized because all remaining birds were in the Luquillo forest, which was heavily damaged in 1989 by a direct hit from Hurricane Hugo. The same hurricane subsequently struck the Francis Marion National Forest, home to one of the largest populations of another endangered bird, the red-cockaded woodpecker (*Picoides borealis*). The Marion Forest is eight times the size of the Luquillo forest, however, and is only one of six major aggregations of the woodpecker, so Hurricane Hugo was not a threat to the persistence of this species (Simberloff 1994).

Therefore, without restoration at large spatial scales, the goal of protecting all species and ecosystems cannot be achieved. The exact sizes of the areas required are sensitive to the climate, species, and communities in the region. In the relatively dry Inland Northwest, for example, fires burned annually up to 2.4 million hectares in an 80-million hectare region prior to 1900, dropping to less than 1.2 million hectares annually since 1900 (Barrett et al. 1997). In the more humid Pacific Northwest (Oregon and Washington), by contrast, 450,000 hectares burn annually on average (Agee 1993).

Acknowledging the importance of thinking about large spatial scales for wildlands restoration does not mean that small-scale restoration projects are unimportant. To some extent, large-scale restoration will have to be promoted by linking multiple small-scale efforts. In short, landscape-scale conservation can achieve its goals only if we "think globally and act locally." Each local project may focus on slightly different priorities and technologies. What is needed, however, is for the many small-scale restoration projects in a region to be carried out under the same conceptual framework, so that elements important to the regional effort are not omitted. The regional perspective leads each local effort to account for its goals—increasing the chance that regionally important species and processes are not ignored because each local project, operating independently, assumes that other local projects will foster this species or that process. Further, certain activities undertaken at a local level might be incompatible with activities at other local sites in the region; control of nonindigenous species offers several examples. Above all, a regional approach does not assume that every square meter of land or water in the region must be restored. Most restoration projects will be implemented as spot treatments. For example, not every riparian area will require the removal of invasive exotics or alteration of its hydrology.

A regional approach also reduces the danger that local restoration projects produce ecological museum pieces—single representatives of communities that, although present because of unusually large restoration and maintenance investments, do not exist in any ecologically meaningful way. Much wetland restoration exemplifies this problem. Single wetland restoration projects may be able to restore some of the traits of an individual wetland. But without restoration of wetlands at many sites throughout a region, essential ecological processes served by a wetlands system—such as control of the regional hydrology or the creation of a patch structure for the persistence of a metapopulation—are lost. Finally, a regional approach should generate economies of scale (Cairns 1993). Many restoration procedures (such as controlled burns) are very expensive on a per-unit-area basis when conducted on a site of a few hectares but are much cheaper over large areas. Initial investments for equipment as well as ongoing stewardship personnel costs are also more efficiently employed when they can be applied to multiple sites.

Just as restoration requires attention to spatial scale, so it demands attention to scales of time. To restore native species and ecosystems, we must consider how distributions of natural communities have changed in the past and how they may change in the future because of alterations in abiotic conditions, whether anthropogenic or not. A local restoration project or protected area may be needed not so much for its current contribution to wildlands conservation as for its role in promoting the persistence or reintroduction of species and communities in the future.

## What Is Needed?

While the general and proximate goals of restoration for regional networks of wildlands (such as those envisioned by TWP) can be fairly easily articulated, what is needed to accomplish these goals? Above all, the social, economic, and political climate must be suitable. If there is not broad support for a restoration project, it is unlikely to succeed. Because regional restoration deals with large areas and large numbers of stake-holders, it is more difficult to achieve this supporting social environment than it is for most local restoration projects—local projects often involve only a single landowner, for example. While the principal methods of ecological restoration will depend on the habitats natural to a region, on the key natural forces influencing ecological communities, and on the various human impacts impinging on these communities, several inter-acting strategies are generally called for. Different ecological communi-ties are governed differently: resources play a key role in some commu-nities and predators in others (Matson and Hunter 1992). These different modes of organization and control dictate different procedures to effect restoration. The classical view of restoration as an acceleration of pri-mary or secondary succession (Bradshaw 1987) perhaps pertains more to communities controlled from the bottom up by resources. In a commu-nity controlled from the top down by predators, by contrast, one can envision a restoration that does not involve mimicking succession. For example, reintroducing a top carnivore to a system may ultimately lead, through great modification of the herbivore trophic level, to change in vegetation and thence to change in the entire community (Peterson and Page 1983), but the trajectory of this change would not be successional.

Restoration methods for wildlands can be divided into three cate-gories: control of invasive nonindigenous species; reestablishment of nat-ural abiotic forces; and reintroduction or augmentation of native species. While small-scale restoration projects often use one or more of these methods, few restoration efforts have considered the interplay of these three factors. The complexity in reclaiming or restoring an acceptable semblance of an ecological community arises from the interplay of abi-otic forces with species interactions that create and sustain natural pat-terns of species numbers and relative abundances over both time and space. The most commonly used restoration method is the introduction or reintroduction of a few plant species—sometimes the formerly domi-nant species in the target community and sometimes fast-growing but not necessarily native species (Cairns 1991; Holl and Cairns 1994). The goal of these efforts is often to speed up succession in an effort to recre-ate quickly the natural physical structure and the gross species patterns of a more natural community (Bradshaw 1987). Depending on the meth-ods used, this strategy alone can often yield rapid changes in an inten-sively degraded site—especially one severely impacted by mining, pollu-tion, or overgrazing. But these short-term gains are often short-lived unless combined with other restoration methods. Furthermore, where large areas are targeted for restoration, and where these areas are degraded as opposed to being sterilized, less conventional restoration methods may be called for, as we shall see.

## Controlling Exotics

First among these other restoration methods is the removal or partial control of aggressive nonindigenous species (Towns et al. 1997). Even in sites with a full complement of native species, nonindigenous species can sometimes distort natural structure and relative abundance patterns, leading to the absolute requirement to control them. Invasive species include plants, animals, fungi, and microorganisms. There is clear evidence that nonindigenous species from each of these groups can alter ecological communities severely—often through indirect, as well as direct, interactions (Simberloff et al. 1997). Thus no large-scale restoration plan is complete without consideration of the array of current invaders and their individual and cumulative impacts on native species. In many instances, with sufficient effort, a nonindigenous species can be controlled at an acceptable level or even eradicated (Simberloff et al. 1997; Towns et al. 1997). For some destructive invaders, however, no effective control method is available or else sociopolitical pressures prevent control. Feral pigs (*Sus scrofa*) that devastate the native vegetation in the mountains of Maui can be adequately controlled for conservation purposes by a combination of methods, including fencing and snares, for example, but opposition has arisen, paradoxically, from both hunters and organizations opposed to hunting animals (Holt 1994). Assuming that the climate of North America warms significantly over the next decades and centuries, the invasion and spread of subtropical aliens (and the diseases they vector) is likely to accelerate—posing a suite of complex problems such as those that now plague Florida (Soulé 1990).

## Reestablishing Abiotic Forces

Second, the operation of natural abiotic forces is clearly necessary for long-term restoration of most natural areas. The most widely understood of these forces are agents of periodic disturbances, including fires and floods. In both cases, numerous studies have shown that normal community structures cannot be maintained in the absence of disturbances (Kirkman and Sharitz 1994; Moreno and Oechel 1994; Minnich et al. 1995). As with nonindigenous species, the action of these forces is not always due to direct impacts (such as mortality) but can also be generated through subtle influences on competitive and trophic interactions (Tyler

1996; Wootton et al. 1996). And there is further complication when considering disturbance factors: what must be "restored" is often a complex regime of intensities and frequencies of events over the proper spatial scales (Mount 1995), as well as the proper biotic community to "receive" the impacts (open-forest understories to carry ground fires versus brushy understories leading to crown fires).

In addition to the well-known disturbance agents of fire and flood, other abiotic forces are now understood to exert strong effects of direct relevance to ecological restoration. Hurricane damage (Horvitz 1997), rare freezes (Steenbergh and Lowe 1977), and geological events, such as soil movement or volcanism (Paine and Levin 1981) strongly influence some natural communities. For these cases, in which the abiotic force is usually uncontrollable, the biggest concern for restoration is not to control the process but to assure that the restoration area is of sufficient size and configuration to allow naturally heterogeneous patchworks of adjacent areas with different times since disturbance (White and Walker 1997).

## Reintroducing Native Species

The final requirement, often not fully considered in restoration programs, is the reestablishment of native species other than dominant plants. These include two vague and overlapping classes of organisms that may have been extirpated or significantly reduced in the restoration area. First are species of plants, animals, and other taxonomic groups that are not known to be strong ecological interactors (Paine 1969; Power et al. 1996) but should be targeted for restoration because of their importance to humans—for symbolic or aesthetic reasons, for example. Second are species that, although not the dominant plants usually targeted in restoration projects, nevertheless play strong roles in determining community structure. These are often called "keystone species," although the exact criteria for this category are debated (Mills et al. 1993). The distinction between the classes is fuzzy because we rarely have direct evidence for or against a species' importance for other members of its community. Nevertheless, there is increasing evidence that many herbivore and predator species, as well as some uncommon plants, do have strong ecological effects such that maintenance of a natural community structure is unlikely or impossible in their absence (Chapter 3; Power et al. 1996). These strong interactors include species whose activities create small-scale disturbance patterns (gophers, Geomyidae; bears, Ursus spp.; badgers, Meles meles and Taxidea taxus; and porcupines, Erethizodontidae), herbivore species that can alter and control the species composition and the physical structure of their communities (beavers, *Castor* spp.; rhinoceroses, Rhinocerotidae; and elephants, Elephantidae), species that serve as important resources for other species even though they are relatively uncommon (figs, *Ficus* spp.; plants that flower in seasons when other plants do not), and finally the classic keystone predators (Paine 1969). In addition to the short-term studies just cited, theoretical arguments have also been made that some animal species—megaherbivores in particular—may have strong effects on community structure over long temporal and spatial scales (Owen-Smith 1987, 1988; Zimov et al. 1996). The subtle impacts of these various classes (and others) of large animals are discussed in Chapter 3.

Although ecological theory and empirical studies now provide numerous examples of the importance of nondominant species in maintaining or recreating natural ecological communities, for most species we do not have clear evidence of importance or interaction strength. In the absence of such information, it is important to take a precautionary view of reintroductions. Too often restoration has excluded consideration of animal species—especially where the goal has been to re-create intact community dynamics. There is a strong ecological argument for the wider reintroduction of many members of original communities that have been extirpated than is usually practiced in restoration efforts.

## A New Paradigm

In sum, we believe that a set of interacting efforts will generally be required to restore large areas to a more natural state. There are numerous interactions between reintroductions, control of nonindigenous species, and reestablishment of abiotic processes that will make successful regional restoration a complex medley of concerted actions. This vision of large-scale wildlands restoration is quite different—and more complex—than the typical approach to small-scale habitat restoration. In fact, it is qualitatively different: a new paradigm. But as we have seen, the scientific data support the view that this medley of efforts is needed to re-create sets of functioning communities of native species that approximate a more natural situation in most areas of the world. The specifics of how to make these methods work in a local or regional restoration effort are much less understood, however.

In the rest of the chapter we review the methods available to enact these restorations, paying particular attention to what is not known and also to interactions between these different methods that will require careful coordination. A general problem in the restoration literature is that most of the empirical tools that have arisen from local projects have not been studied in controlled experiments. Multiple factors are not distinguished in separate treatments, replication is often poor or nonexistent, and the scale of the effort is so small that it is unclear whether similar results will be obtained on larger scales.

Restoring large networks of habitats provides an opportunity to carry out large-scale, experimental studies that could inform future restoration efforts as well as basic theory in community and ecosystem ecology. Further, such efforts can be made compatible with plans for the adaptive management of protected areas. For example, well-replicated, factorial designs that separate physical from biotic treatments can be used to discover which elements are critical to successful restoration of species or ecosystem attributes in differing situations. Miller et al. (1983) examined both the physical and biotic effects of topsoil stockpiling on revegetation success in semiarid rangelands and discovered a complex interaction involving the promotion of grasses by the accumulation of snow by *Atriplex* shrubs that, in turn, fostered mycorrhizal establishment. A nonexperimental approach yields only correlative patterns between species occurrences and soil treatments. The well-designed experiments of Miller and his colleagues, however, have advanced our understanding of establishment processes.

## Controlling Exotic Species

Exotic species are a pervasive and often highly damaging feature of contemporary landscapes. A dominant plant like melaleuca (*Melaleuca quinquenervia*) in Florida (Simberloff et al. 1997) or tamarisk (salt cedar; *Tamarix* spp.) in the American Southwest (Loope and Sanchez 1988; U.S. Congress 1993) can so completely change the habitat of a region that virtually all original inhabitants, both plant and animal, are locally extirpated. Animals that devastate native dominant plants, such as feral hogs (U.S. Congress 1993; Holt 1994), can also totally modify a local habitat. The very presence of nonindigenous species, whether or not they harm native species, degrades the wildness of an area in the perception of many. Ideally we would advocate the removal of all nonindigenous species in all areas as a necessary step in regional restoration.

But nonindigenous species are so ubiquitous that their complete elimination from most areas is unlikely. First, for many species, particularly plants and insects, we do not have the technology to ensure elimination, although maintenance control at some environmentally acceptable level is often feasible (Simberloff 1996; Simberloff et al. 1997). Second, there is often widespread opposition to the elimination of introduced species, even when they can be shown to be ecologically damaging. Animal rights advocates object to removal of vertebrates when the only feasible methods entail killing them (such as the monk parakeet, *Myiopsitta monachus;* Faber 1973). Hunters often wish to maintain populations of nonindigenous game species (such as feral hogs; Gustaitis and McGrath 1991; Tanji 1993). Even a damaging introduced dominant plant, such as eucalyptus (*Eucalyptus* spp.), finds passionate defenders when a removal project is proposed (Westman 1990). Third, with our increasingly mobile societies and global economies, the movement of species across habitat, state, and national boundaries is one process that will continue into the future—making it impossible to prevent their establishment and reestablishment in natural areas (Schmitz and Simberloff 1997).

Most exotics, however, have few known effects on native species or on ecosystem functions; indeed, only a small minority is known to cause extensive damage to natural communities (Williamson 1996). Therefore, we can focus our efforts on controlling only those species that have serious effects or cause particular problems—and only in the habitats or regions where they are truly problematic (unless a local control effort will be continually defeated by reinvasion from other populations within the region). Of course, many nonindigenous species have become problematic only after a long time lag (Crooks and Soulé 1996). To discuss the steps needed in evaluating control of nonindigenous species for large-scale, wildlands restoration, we break down our discussion into habitat-specific and regional components of assessment and control and then consider the technologies currently in use.

## Habitat-Specific Aspects of Control

The first goal is to develop priorities for action based on thresholds of damage. While such priorities will be system-specific and species-specific, it is important to recognize that invasive species have numerous impacts: they may eat other species; they may be pathogens; they may transmit diseases to other species; they may poison other species; they may alter hydrological, fire, or other natural cycles; they may compete with native species for food, space, or pollinators; they may interfere with other species without actually taking a resource, as in aggressive behavior of animals or allelopathy of plants; and they may hybridize with native species. While these various effects are crucial in ranking the ecological importance of different nonindigenous species, the literature on biological invasions suggests that we cannot determine from life-history traits which species are most likely to have significant impacts on native communities (Williamson 1996). Thus we shall not pursue such categorization further—except to note that it may not always be obvious which nonindigenous species are most important. Although purple loosestrife is a highly visible invader, the evidence so far is weak that it has had significant population impacts on any native species in North America (Hager and McCoy 1998). There is no substitute for direct field study of the impact of a nonindigenous species on natives.

Hard on the heels of establishing such priorities is the question of what control methods are either feasible or necessary. Often the presence of a nonindigenous species prevents the establishment of native species: in this case, restoration will require its removal. Nonindigenous species may alter ecosystem processes. Thus, for example, it is impossible to reestablish native vegetation in wetlands that have been drained by salt cedar without removing the salt cedar. Sometimes the impact of a nonindigenous species on a native one (such as eating it) will preclude reestablishment of the native. Continued grazing by cattle or sheep, for instance, may prevent the establishment of native bunchgrasses in parts of the American West. But in all these instances it may require only a temporary or permanent reduction of numbers—rather than complete elimination—to achieve adequate control for the purposes of wildlands restoration. Reduction of a fire-conducting nonindigenous weed, for example, may suffice to give native vegetation a chance to establish sufficiently that the weed cannot invade in densities that would propagate damaging fires. How much removal is needed and for how long? These questions can be answered only through controlled field experiments.

As there is an extensive literature on methods for controlling exotics (U.S. Congress 1993; Cronk and Fuller 1995; Simberloff et al. 1997), we need not summarize it here. We must, however, reemphasize that the restoration of disturbance regimes or the reintroduction of certain native species will set up the conditions necessary for long-term suppression of certain nonindigenous species and that these projects will perforce be regional. For example, an appropriate burning regime in longleaf pine (*Pinus palustris*) ecosystems will kill seeds and adults of many invaders; the return of seasonal flooding in the Everglades will drown many weedy nonindigenous species; and the restoration of natural flooding on rivers will allow the flushing or scouring of riverbank communities, removing many nonnatives. Beavers can be used to aid the reestablishment of riverine communities because their dams will raise the water table, drowning nonindigenous grasses and shrubs and favoring native willows. Reintroduction of wolves and other carnivores may help control herbivore populations; when uncontrolled, the herbivores may prevent the reestablishment of native, palatable species while favoring the spread of unpalatable introduced species. In general, a probable consequence of landscape-scale restoration efforts will be the development of communities that are less vulnerable to invasion (via removal of unnatural disturbance that facilitates invasion)—although the reasons why some communities are more easily invaded than others is an area under active study.

## Regional Aspects of Control

If the ultimate goal is regional restoration, as in The Wildlands Project, it is important to evaluate procedures for local control of nonindigenous species within a regional context. Some harmful invaders disperse well and reproduce rapidly; thus they would have to be controlled everywhere to achieve effective control anywhere. It would hardly pay to undertake an expensive local project to remove leafy spurge (*Euphorbia esula*) or musk thistle (*Carduus nutans*) if adjacent lands provide seed sources for immediate recolonization. At the other extreme, species with low dispersal abilities and low population growth rates could be feasibly targeted for intensive local control. Venus flytrap (*Dionaea muscipula*) has been established as several small populations in the Apalachicola National Forest in northern Florida; these are barely stable, and no new populations have been found. As the nearest other populations are in the Carolinas, it would be feasible to control or even to eradicate the Florida populations if they were deemed damaging.

A number of intermediate situations are possible. For example, an introduced population of a highly dispersive nonindigenous species may be so isolated from other populations that a local eradication of the former need not be futile if the latter populations persist. Or a species that is inherently nondispersive may be the target of continued illegal reintroductions that render its local control impossible. One way to judge the reintroduction of native species and natural processes is in their effect on invasive species. To the extent that reintroductions can suppress the demographic potential even of highly dispersive invaders, they may render local control efforts feasible.

Although intensive control of nonindigenous species would ideally be restricted to local sites in the core areas—with a lower degree of control in the buffers and little or none outside lands designated for protection—such an approach may be futile. It all depends on the biology of the species and the configuration of designated lands and potential source areas for propagules. It may be that consortia of organizations and authorities will have to cooperate in the control of certain nonindigenous species throughout an entire region.

## Technologies for Control

Broadly speaking, there are three general approaches to removing nonindigenous species: mechanical, chemical, and biological control (Simberloff et al. 1997; Simberloff in press). Although each approach is useful in specific situations, none is a silver bullet. Moreover, all may be ineffective or have detrimental effects regionally even if they are locally effective. It is possible to control certain plants and animal species at acceptable levels locally simply by removing them by hand. This method is extremely labor-intensive, however, and not feasible over very large areas. The exact limits to such an approach depend, of course, on the case. Indeed, various social and technical advances may make mechanical control feasible over surprisingly large areas. Mechanical reapers have largely controlled water hyacinth in Florida, for example, where it had been an ecological scourge (Simberloff et al. 1997). The Nature Conservancy has had great success using volunteers to remove nonindigenous plants locally (Randall et al. 1997). And as more states pass legislation requiring convicts to work, it is possible that the labor pool will increase greatly. In Oregon convicts are already manufacturing gypsy moth (*Lymantria dispar*) traps; in Florida they are hand-cutting melaleuca and Brazilian pepper trees (*Schinus terebinthifolius*).

Chemical control is often promoted as an alternative when mechanical control is not feasible. Although the widely publicized human health consequences and population impacts on nontarget species of such early pesticides as DDT have made many conservationists chemophobic (Williams 1997), certain newer pesticides are sufficiently free of these problems that their use is even endorsed by, for example, the Worldwide Fund for Nature (Cronk and Fuller 1995) and they may be extremely effective locally (although cumulative effects may be manifested in the long term). Two related problems arise, however. First, chemicals are often so expensive that a regional application is economically prohibitive. Second, target organisms often evolve resistance, so that greater and greater amounts of a pesticide have to be applied—not only exacerbating the economic problem but increasing the potential for unintended ecological side-effects.

Biological control is widely viewed as an inexpensive "green" method of controlling nonindigenous species. Sometimes it works very well. But only a minority of biological control introductions actually confer economically significant control (Williamson 1996), while there is increasing concern about impacts on nontarget species (Simberloff and Stiling 1996). *Rhinocyllus conicus,* for example, a European weevil introduced to North America for control of Eurasian musk thistle in pastures, has been found attacking native thistle species, including certain narrowly restricted endemics, in three national parks and two Nature Conservancy preserves far from the sites of release (Louda et al. 1997). Because biological control organisms are alive, they reproduce, evolve, and disperse autonomously. Thus local use to attack a nonindigenous pest in one site may lead to regional, national, or even international spread and attack on nontarget species.

The challenge of nonindigenous species control posed by the regional and national scope of conservation networks is enormous. Further, true ecological restoration is often stymied by these invaders and restoration ecology cannot advance without much more research on this topic (Randall et al. 1997; Towns et al. 1997). To the extent that we are able to control nonindigenous species, it will contribute greatly, not only to the success of regional wildlands conservation projects, but to the science of restoration ecology.

## Restoring Native Species

Reintroductions for large-scale restoration projects have three broad goals: to restore the original structure and relative abundances of dominant species; to reintroduce species that are often rare and of little-known overall importance to ecosystem function; and to provide services for other species of critical importance (as forage, cover, and so forth). Because these three types of reintroductions arise from different traditions in applied ecology and sometimes use different techniques, we discuss each in turn.

### Restoring Working Communities

The reestablishment of thriving populations of native plant species forms the backbone of traditional restoration ecology (Bradshaw and Chadwick 1980). As we have seen, the historical focus of these efforts has been to re-create a functioning ecological community on heavily polluted and denuded sites (Bradshaw 1987). Note that this goal is not necessarily a part of restoration as we have formally defined it, as it may entail replacing the native community. To the extent that the emphasis is on ecological processes and functions rather than species, such efforts are unlikely to contribute to restoration in the strict sense. Because of this historical focus, the greatest expertise in restoration ecology has been in the interaction of soils with plants and the reestablishment of relatively few species per location. The roles of plant reintroductions in regional restoration projects as envisioned here differ somewhat from these historical efforts. Above all, the target areas will probably be sites in which there is an extant community of native species that must be changed— not re-created—to become more natural or to regain certain species or community characteristics. Moreover, animal reintroductions will play a key role in rewilding approaches such as those espoused by TWP restorations (Soulé and Noss 1998). Reintroduction or augmentation of a keystone animal species such as the beaver, for example, may govern an entire local or even regional restoration project. Nonetheless, lessons from traditional restoration ecology as well as basic population and community ecology can aid regional restoration efforts.

First, restoration ecology has adapted horticultural techniques to provide fairly reliable methods to reestablish certain plant species. These methods include site preparation, outplanting, outseeding, seed dormancy manipulation, and clonal propagation. Perhaps more an art than a science, this basic knowledge has been hard won and lays the groundwork for successful restoration efforts involving many kinds of reintroductions or population augmentations. This work will have to be expanded to achieve restoration on very large scales—for most projects to date have restored only small areas (10 to 100 hectares). The Kissimmee River restoration project (Toth 1995), 11,000 hectares, is one of the largest to date.

Second, particular disturbance patterns, such as specific hydrological or fire regimes, are often required for population establishment. Traditionally the emphasis has been on community stability, but the principle is equally valid for population persistence. In fact, if one emphasizes species' identities rather than ecological processes in defining a community, community stability is probably just an epiphenomenon of population persistence (Shrader-Frechette and McCoy 1993).

Third, restoration ecology has shown the importance of establishing sufficient abundances or densities of individual plants in order for successful reintroduction to occur. Similar information comes from the literature on introduced species—for example, an important predictor of successful establishment for insects introduced for biological control (Beirne 1975) and birds introduced to New Zealand (Veltman et al. 1996; Duncan 1997) is the number of individuals introduced. For bird and mammal translocations, Griffith et al. (1989) portray a strong relationship between probability of establishment and number of individuals released. The reasons for this empirical result are seldom examined in a restoration context, but there are several possible explanations: demographic stochasticity (Lande 1987, 1993; Menges 1992); inbreeding depression owing to small population sizes (Barrett and Kohn 1991; Huenneke 1991); problems with pollinator or seed disperser attraction at small population sizes (Woods 1984; Kunin 1993; Groom 1998); establishment limitations owing to inadequate mycorrhizal densities that can occur when densities of conspecific or key heterospecific plants are insufficient (Miller et al. 1983; Miller 1987); inability of newly reintroduced low-density populations to shade out or otherwise resist invasion by invasive native or exotic species (Mehrhoff 1996); density effects in preventing soil erosion or nutrient leaching problems (Bradshaw and Chadwick 1980); and breeding displays and other means by which high density may facilitate animal breeding (Allee et al. 1949; Simberloff 1986). Whatever the relative importance of these mechanisms, there is clear evidence of strong Allee effects for many species (Dennis 1989; Lamont et al. 1993; Groom 1998). This result is important for large-scale restoration. For example, rarely are dominant plant species truly extirpated from large regions in which we hope to reestablish them. But their former abundances are often so reduced that population augmentation may be crucial to restore more natural relative abundances; populations will not increase sufficiently on their own.

And fourth, experimental community ecology suggests that strong priority rules sometimes govern community assembly. That is, the final community structure (its composition and relative abundances) that arises from a set of species may be highly sensitive to the order in which they are introduced (Simberloff 1990; Samuels and Drake 1997). Although this idea gets its strongest support from microcosm studies, there are clear examples of this phenomenon from applied and basic animal and plant ecology. Obviously this knowledge has important implications for restoration. But thus far there are no dependable rules to determine the exact consequences of different reintroduction orders—beyond obvious imperatives such as reintroducing a host plant before reintroducing its host-specific herbivore. In fact, the literature on restoration and the literature on assembly rules—though clearly dealing largely with the very same topic—have been remarkably independent. Samuels and Lockwood (1998) have recently attempted to draw lessons from community assembly study for restoration ecologists. They find that, so far, the assembly rule literature does not so much dictate specific methods in restoration as caution that large-scale reintroductions should be preceded by small-scale pilot studies to avoid wasting time and effort.

Phytoremediation is a new technique that may greatly aid restoration of certain sites (Dobson et al. 1997). Many plants are adapted to habitats in which trace elements essential to their development are present in very low concentrations. A number of plant species have thus evolved highly efficient mechanisms for removing trace elements from the soil. As these trace elements are often present in significant concentrations as pollutants of abandoned industrial wastelands, plants may be used to clean the soil (Baker and Brooks 1989). In some cases, it may even be possible to harvest the plants and reclaim the trace elements from them for profit or at least to cover the cost of the remediation (Baker et al. 1994; Reeves et al. 1995). This technique is particularly promising for severely degraded land, such as Superfund sites, that would otherwise be unusable. It would be wildly optimistic to think that such land could be restored to sufficiently "natural" condition as to become a core reserve, but it might become useful as a friendly buffer or even provide connectivity among core preserves. Further, such reclaimed land could almost certainly be converted to human recreational use and thus help to finance the pristine parts of a regional reserve network. As the plants used in such phytoremediation are generally nonindigenous, their use in a restoration project would entail plans to limit their density and spread, if not to eradicate them once they have served their remediation role.

## Reviving Natural Species Diversity

Many reintroductions during a regional restoration project may involve species with little or no known importance to other community members. In particular, rare and endangered species are likely to need true reintroductions because they have been completely eliminated from large areas of potential habitat. In these cases, restoration must pay even more attention to the possibility of Allee effects. For one thing, the propagules to be used in the reintroduction may be difficult to come by—in some instances, in fact, they may constitute an entire species population. Although the reasons why some species are common and others are rare are often unclear (Gaston 1994), many rare species have traits that make extinction more likely (Gaston 1994; Kunin and Gaston 1997). Further, the size of the number of propagules available for reintroduction of a rare species is likely to be smaller than that for a common species—as the literature shows—small propagule number is strongly associated with failure. Because of the lower likelihood of success and the value of the propagules, reintroductions of rare species should usually be performed after the reestablishment of disturbance regimes and healthy populations of many dominant species. Many past conservation efforts to augment and reintroduce rare plants have shown that reintroduction in the absence of substantial efforts to improve habitat quality are futile. Finally, while almost all reintroduction efforts have targeted free-living plant or animal species, we must recognize that fungi, bacteria, viruses, and assorted other pathogens and symbionts are parts of all communities. Hence reintroduction of such species should be part of regional conservation efforts. Perhaps most important, such species may play crucial but poorly recognized roles in maintaining community structure.

## Providing Services for Other Species

The reestablishment of other interacting species—especially pollinators and mycorrhizal symbionts (Miller 1987)—should be considered part and parcel of rare plant reintroductions. Moreover, forest and prairie restorations have sometimes been impeded by the absence of natural densities of ants, which disperse some plant species and also generate small-scale disturbances that may be crucial to the establishment of certain plants (Woods 1984; Kline and Howell 1987). The successful reintroduction of animals to a landscape may require the establishment or augmentation of critical food plants, animal prey, cover species, or other elements. Reintroduction for this purpose—assisting the survival of a prominent target species—may use many of the methods outlined earlier. But depending on the rarity of the interacting species to be fostered, it may be necessary to perform laborious experimental research to determine the exact requirements for persistence. Highly specific disturbance regimes may be necessary, for example, or the specific spatial distribution of certain plant species relative to the expected movements and home range use patterns of a target animal species may be crucial.

Restoration at the landscape scale seeks to reestablish the same conditions that determine success in the most ambitious localized site-based projects: a specified composition, structure, and function for the system. As we have seen, the traditional approach in restoration is to create the gross species composition by planting, translocating, or attracting plant and animal species specified by regulatory agencies—arguably the easiest part of restoration. This is typically a short list of readily acquired dominant or indicator species devoid of requirements for nonvascular flora, microorganisms, or soil fungi, let alone animals. Re-creating a desirable structure for the site is much more difficult. Varying the ages and sizes of specimens, their placement within the site, and the nature of the substrate into which they are introduced—all can be used to achieve an appropriate structure. But there is one crucial ingredient affecting structure and composition that cannot be readily introduced: time. It is usually impractical to introduce old-growth trees, viable populations with an adequate prey base, and developed soil.

Time is also critical to reestablishing the last, most important, and most difficult aspect of systems restoration: function. The goal of restoration includes re-creating the functional attributes of the destroyed system so that it can regain its role in regional landscape integrity. Certain physical functions can be re-created quickly: wetland basin morphology can be sized to store appropriate volumes of water; planting of terrestrial vegetation or application of mulch can reduce or prevent soil erosion; initial net primary productivity can form the basis for a simple food web; added fertilizer can jump-start nutrient cycles. But the restoration of desirable water quality characteristics, old-growth conditions necessary to support certain species, and complex systems of energy and nutrient transfer can be achieved only over extended periods. Composition, structure, and function interact in complex ways.

Reintroduction and maintenance of natural biotic and abiotic processes that govern systems are essential for full restoration. Otherwise a restoration may be impossible or, at best, expensive artificial means will be required to sustain desired characteristics. The Coachella Valley fringe-toed lizard (*Uma inornata*) persisted in dwindling sand-dune habitats in southern California, along with other dune-adapted species. A habitat conservation plan was developed to preserve 7000 hectares of suitable habitat and montane sources of aeolian sand necessary to nourish the dunes (Committee on Scientific Issues 1995). The initial reserve configuration was inadequate, however, partly because of insufficient knowledge of the precise sources of the sand and partly because railroad rights-of-way and associated planted trees inhibited sand movement while flood-control ditches trapped fine materials from which the sand develops. This problem has yet to be solved (Bean et al. 1991).

What processes are critical for restoration, why are they important, and how can they be restored? Numerous ecological, hydrological, geological, morphological, astronomical, and evolutionary processes affect the distribution of species and integrity of ecosystems. Among the primary physical processes are continental drift, volcanism, earthquakes, climate variability, weather events, glacial cycles, meteor/asteroid impacts, and solar flares. Generally these are large-scale physical processes operating over long periods and subject to little short-term influence by biological factors or by humans. Some physical processes, however, are greatly influenced by biological factors and humans: soil profile development, erosion and sedimentation, hydrological cycles, nutrient cycles, and fire. Biological processes with variable influence from physical factors include evolution, succession, predation, herbivory, species migration and invasion, decomposition, pollination, and parasitism. A main objective of reserve design should be the maintenance of these processes (Noss 1992). The natural patterns of distribution and abundance of biota are largely determined by the response of that biota to historical cycles, processes, and disturbance regimes. Where these historic determinants have been altered, they must be restored if the system is to be maintained.

Most alteration of process has been deliberate. For example, wildfire was suppressed in order to protect trees and property. Free-flowing rivers were dammed to prevent flooding and allow changes in floodplain use. Large predators were controlled to minimize losses of livestock. Natural succession was halted or reversed to increase yields of selected products. Erosion was allowed to increase to aid cultivation of certain crops. Each of these changes may have accomplished someone's immediate objective, but the long-term impacts on biodiversity were either unknown or not considered—and the cumulative impacts often extended well beyond the vicinity of implementation.

It is impossible to say which processes are most important in restoration. Many are beyond our ability to manipulate. Those that can be controlled or reintroduced at reasonable cost and within appropriate time frames are the only ones that should be seriously considered by managers. From the perspective of landscape-level or regional efforts, certain processes should be seen as priorities and addressed in detail:

- Processes that directly affect the objectives—rewilding, protecting biodiversity, and maintaining ecosystem integrity in the case of TWP
- Processes that operate within the project's time frame
- Processes that have significant short-term negative impacts when altered
- Processes that respond to effective restoration and manipulation
- Processes that operate at the landscape level

Such processes would include predation, fire, nutrient cycles, species invasion and migration, succession, hydrology, pollination, erosion, and sedimentation.

Any restoration of process must be based on past, present, and future anticipated conditions, and a baseline model will be needed against which the response can be compared to determine success. The ability to restore disrupted processes may be limited by lack of knowledge, insufficient skill, paucity of resources, and politics. But doubts about implementation should not preclude addressing the need for process restoration in the planning and design phases of conservation projects. Relative frequency, intensity, duration, extent, and other variables must be articulated so that data and logistic needs can be assessed.

Restoration of ecological processes disrupted by humans often engenders opposition from certain segments of society or encounters unfavorable political and regulatory climates. The most controversial—and probably the most essential from an ecological point of view—are fire, hydrology, and predation. We have already discussed restoring predation regimes. Extensive fire-dependent ecosystems once covered much of the southeastern, midwestern, and western United States. Healthy longleaf pine or ponderosa pine (*P. ponderosa*) forests and oak savannas require frequent, low-intensity ground fire (Platt et al. 1988; Covington and Moore 1994; Faber-Langendoen and Davis 1995). Florida sand pine (*P. clausa*) scrub requires irregular, high-intensity, stand-replacement fire (Christman 1988). Fire suppression ultimately results in the replacement of these communities and loss of large suites of rare and endemic taxa. The reintroduction of fire through prescribed burning has been a slow process and is hampered by regulatory control for air pollution, lack of "right to prescribed burning" legislation in most states (Florida enacted such a law in 1991), and public antipathy—partly a residue of the old Smokey the Bear attitude that all fire is bad and must be prevented (Pyne 1982; Pyne et al. 1996).

Prescribed burning is an alternative to natural wildfire ignition. Although it is costly, subject to managerial changes in attitude and priority, and not always representative of natural fire, there is a well-developed technology for propagation and management of fires in various plant communities (USDA Forest Service 1989). Allowing completely natural burns may hinder restoration, so a combination of prescribed burning and fire suppression may be essential during the restoration process. As with species reintroductions, successful restoration of fire and other processes often requires extensive experimentation until we know that the size, physical structure, and biota of the restoration area are certain to be aided rather than damaged by fully natural disturbance regimes.

Widely fluctuating water levels (including flooding and low flow) are critical components of many ecosystems. Today, however, dams, channels, levees, and streambank armoring have massively modified the flow regimes of many watercourses. Floodwaters flush sediment and silt from streambeds valuable for fish spawning, deposit sediments and nutrients on floodplain terraces, create and renew ponds, wetlands, and marshes used by many species not found in the mainstream channel, produce food for waterfowl whose migrations are timed to natural flood cycles, scour and redeposit organic debris that is the basis for extensive detrital food chains in floodplains and estuaries, and help to create islands and channel complexity by shifting and trapping silt and downed trees (Mount 1995; Robbins 1998). Moreover, low water periods are important for oxidizing organic sediments, concentrating organisms in low water pools for specialized feeding by species such as wood storks, and providing an appropriate seedbed for semiaquatic plants that cannot root in open water (Wharton et al. 1977). Many native species have evolved adaptations to the natural ebb and flow of a river.

Restoring flood regimes is controversial for some of the same reasons that controlled burns are controversial. Dams, levees, and channelization allow new land uses and construction on former floodplains. Reservoirs and channelization not only permit new forms and areas of recreation, transportation, and irrigation but provide water supplies to cities. The interests that fostered the changes in water flow and floodplain dynamics in the first place often oppose efforts to redress the damage to natural system—seven when the fact that damage has occurred is not in dispute (Mount 1995). Reestablishing key flooding conditions can be as simple as releasing peak flows from a dam—as in experimental releases from the Glen Canyon Dam to re-create flood flows in the Colorado River through the Grand Canyon in 1996 (Wegner 1997). But the restoration of natural flood regimes is often a massive undertaking requiring removal of water control structures, extensive earth moving, complex hydrological modeling, much on-the-job experimentation, and lots of time and money.

In Florida, for example, restoration of the Kissimmee River was first discussed less than three years after the channelization and water control public works project was completed in 1971. At a cost of some $55 million, a 160-kilometer meandering river was converted to an 85-kilometer canal and 16,450 hectares of dynamic floodplain wetlands was reduced to 3550 hectares of impounded wetlands behind five water control structures (Pruitt and Gatewood 1976). Waterfowl and wading bird populations were reduced by 90 percent. By 1981, a plan for partial restoration had been developed and was moving through the approval and design process. Experimental restoration began in the early 1990s, and by 1995 physical restoration of the river channel and removal of structures was initiated. In about a decade, roughly two-thirds of the original river and floodplain will be fully restored at a cost of about $475 million. Initial results indicate that simply re-creating the seasonal extremes of flood and drought is reestablishing the dynamic floodplain ecosystem with little more than human input (Toth 1995). Another current large restoration is the project to restore a major part of the natural flow regime to the Snake River in Wyoming (Robbins 1998). This project entails experimental technologies such as pounding many logs into a riverbed at various sites to trap sediment and re-create islands, as well as excavation of channels in an effort to re-create the anastomosing, braided channels and reduce the energy of current flows in the few remaining channels.

It is imperative that the impacts of these and other disruptions—as well as the technologies for redressing them—be thoroughly investigated. Without knowledge of both the natural processes and any substitutes that humans develop to restore or replace them, it will be impossible to determine the precise impact of their integration into a landscape management program. But it is already clear that without the restoration of processes, substantial ecosystem restoration at the landscape level will not be achieved.

## Overcoming the Obstacles

One goal of The Wildlands Project—to maintain the full range of native species and ecosystems—is so ambitious that the science of restoration ecology cannot yet give us adequate guidance. Although it provides numerous useful techniques, particularly for reintroduction and augmentation of particular species, the scope and scale of such regional/continental projects are beyond those of even the most thorough restorations in the literature—for example, those of small prairies and islands. Even so, such large-scale restoration projects can be a major factor in the current burst of activity attempting to put restoration ecology on a sound scientific footing. So long as project-as-experiment is not just a catchphrase but a modus operandi—complete with appropriate replication, controls, and quantitative analysis—such efforts can accelerate. The fact that regional restoration will undoubtedly have local and even regional setbacks must not be discouraging. The magnitude and novelty of what we are trying to do ensure that success will be neither immediate nor global. Yet crucial benefits will derive even from temporary failures if projects are conceived as scientific hypotheses and we learn from the ones that turn out to be wrong. Regional restoration projects can also lead in the integration of relevant academic disciplines into restoration ecology. Invasion biology, community assembly, community trophic governance, keystone species, and many other areas of ecological endeavor that have been the province of academic ecologists are all pertinent to restoration. Conservation biologists and conservationists must exploit all of them.

Beyond the scientific impediments to restoration, social and legal impediments abound. Earlier we noted the sociopolitical objections to removal of nonindigenous species that prevent restoration. The reintroduction of certain species can be equally controversial: reintroductions of the red wolf in the Southeast and gray wolf and bison in the West have generated considerable opposition. As well, we noted opposition to reintroducing various natural processes (such as fire and floods). Laws often hinder restoration. Regulations on fires near airports or cities, for example, can necessitate enormously costly alternatives or even prevent a restoration altogether. Laws specifying the amount of topsoil that must be established can be just as irrational from a restoration standpoint. Following surface mining, compliance with restoration regulations often leads to areas with reduced topographic and floristic diversity (Steele and Grant 1982). For example, regulations requiring soil to be spread evenly throughout a reclamation project can turn a diverse topography with relief, rock piles, and other aspects of complexity into a far less diverse landscape. As we have seen, many activities colloquially termed "restoration" have different goals than the ones we espouse in this chapter. For instance, establishment of some vegetation on a site or "improvement" in terms of cattle and game productivity are frequently targets. To the extent that advocates of wildland restoration can articulate their goals, rationalize their methods, and enlist public support for their endeavor, they can help restoration ecologists to overcome these barriers of psychology and law.

For the benefits of regional restoration experience to be maximized, they must be propagated. This means that relevant aspects of the projects should be published in the scientific literature and lay media, as well as publicized widely in scientific and public presentations. Restoration ecology has a low profile in scientific circles today—for example, there are few university courses in restoration and not many training programs in restoration techniques. The Wildlands Project should undertake to raise both academic and public consciousness about restoration.

# References

Agee, J. K. 1993. *Fire ecology of Pacific Northwest forests.* Washington, D.C.: Island Press.

Allee, W. C., A. E. Emerson, and O. Park. 1949. *Principles of animal ecology.* Philadelphia: Saunders.

Armstrong, J. K. 1993. Restoration of function or diversity? In D. A. Saunders, R. J. Hobbs, and P. R. Ehrlich (eds.), *Nature conservation 3: The reconstruction of fragmented ecosystems.* Chipping Norton, New South Wales: Surrey Beatty & Sons.

Atkinson, I. A. E. 1988. Presidential address: Opportunities for ecological restoration. *New Zealand Journal of Ecology* 11:1–12.

———. 1990. Ecological restoration on islands: Prerequisites for success. In D. R. Towns, C. H. Daugherty, and I. A. E. Atkinson (eds.), *Ecological restoration of New Zealand islands.* Wellington: New Zealand Department of Conservation.

Baker, A. J. M., and R. R. Brooks. 1989. Terrestrial higher plants which hyperaccumulate metallic elements—a review of their distribution, ecology, and phytochemistry. *Biorecovery* 1:86–126.

Baker, A. J. M., S. P. McGrath, C. M. D. Sidoli, and R. D. Reeves. 1994. The possibility of *in situ* heavy metal decontamination of polluted soils using crops of metal accumulating plants. *Resources, Conservation, and Recycling* 11: 41–49.

Balling, R. C., Jr., G. A. Meyer, and S. G. Wells. 1992. Relation of surface climate and burned area in Yellowstone National Park. *Agricultural and Forest Meteorology* 60:285–293.

Barrett, S. C. H., and J. R. Kohn. 1991. Genetic and evolutionary consequences of small population size in plants: Implications for conservation. In D. A. Falk and K. E. Holsinger (eds.), *Genetics and conservation of rare plants.* Oxford: Oxford University Press.

Barrett, S. W., S. F. Arno, and J. P. Menakis. 1997. *Fire episodes in the inland northwest (1540–1940) based on fire history data.* General technical report INTGTR-370. Portland, Ore.: USDA Forest Service.

Bean, M. J., S. G. Fitzgerald, and M. A. O'Connell. 1991. *Reconciling conflicts under the Endangered Species Act: The habitat conservation planning experience.* Washington, D.C.: World Wildlife Fund.

Beirne, B. P. 1975. Biological control attempts by introductions against pest insects in the field in Canada. *Canadian Entomologist* 107:225–236.

Belous, R. 1984. Restoration among the redwoods. *Restoration and Management Notes* 2:57–65.

Bradshaw, A. D. 1987. The reclamation of derelict land and the ecology of ecosystems. In W. R. Jordan III, M. E. Gilpin, and J. D. Aber (eds.), *Restoration ecology.* Cambridge: Cambridge University Press.

Bradshaw, A. D., and M. J. Chadwick. 1980. *The restoration of land.* Oxford: Blackwell.

Broome, S. W., E. D. Seneca, and W. W. Woodhouse Jr. 1988. Tidal marsh restoration. *Aquatic Botany* 32:1–22.

Cairns, J., Jr. 1991. The status of the theoretical and applied science of restoration ecology. *Environmental Professional* 13:1–9.

———. 1993. Ecological restoration: Replenishing our national and global ecological capital. In D. A. Saunders, R. J. Hobbs, and P. R. Ehrlich (eds.), *Nature conservation 3: The reconstruction of fragmented ecosystems.* Chipping Norton, New South Wales: Surrey Beatty & Sons.

Chapman, R., and A. Younger. 1995. The establishment and maintenance of species-rich grassland on a reclaimed opencast coal site. *Restoration ecology* 3:39–50.

Christman, S. P. 1988. *Endemism and Florida's interior sand pine scrub.* Final report GFC-81-101. Tallahassee: Florida Game and Fresh Water Fish Commission.

Committee on Scientific Issues in the Endangered Species Act. 1995. *Science and the Endangered Species Act.* Washington, D.C.: National Academy Press.

Covert, C. J. 1990. Revegetation of abandoned acid coal mine spoil in south central Iowa. In J. J. Burger (ed.), *Environmental restoration: Science and strategies for restoring the earth.* Washington, D.C.: Island Press.

Covington, W. W., and M. M. Moore. 1994. Postsettlement changes in natural fire regimes and forest structure: Ecological restoration of old-growth ponderosa pine forests. *Journal of Sustainable Forestry* 2:153–181.

Cronk, Q. C. B., and J. L. Fuller. 1995. *Plant invaders.* London: Chapman & Hall.

Crooks, J., and M. E. Soulé. 1996. Lag times in population explosions of invasive species: Causes and implications. In O. T. Sandlund, P. J. Schei, and Å. Viken (eds.), *Proceedings of the Norway/UN Conference on Alien Species.* Trondheim: Directorate for Nature Management and Norwegian Institute for Nature Research.

Daniels, W. L., and C. E. Zipper. 1988. Improving coal surface mine reclamation in the central Appalachian region. In J. Cairns Jr. (ed.), *Rehabilitating damaged ecosystems.* Vol. 1. Boca Raton, Fla.: CRC Press.

Davis, M. B. 1993. *Old growth in the east.* Richmond, Vt.: Cenozoic Society.

Dennis, B. 1989. Allee effects: Population growth, critical density, and the chance of extinction. *Natural Resource Modeling* 3:481–538.

Dobson, A. P., A. D. Bradshaw, and A. J. M. Baker. 1997. Hopes for the future: Restoration ecology and conservation biology. *Science* 277:515–521.

Duncan, R. P. 1997. The role of competition and introduction effort in the success of passeriform birds introduced to New Zealand. *American Naturalist* 149:903–915.

Faber, H. 1973. State acts to wipe out monk parakeet. *New York Times,* 7 April, pp. 1 and 41.

Faber-Langendoen, D., and M. Davis. 1995. Fire frequency and its effect on woody plant cover at Allison Savanna, east-central Minnesota. *Natural Areas Journal* 15:319–328.

Forbes, B. C. 1993. Small-scale wetland restoration in the high arctic: A long-term perspective. *Restoration Ecology* 1:59–68.

Foster, D. R., J. D. Aber, J. M. Melillo, R. D. Bowden, and F. A. Bazzaz. 1997. Forest response to disturbance and anthropogenic stress. *BioScience* 47:437–445.

Fritts, S. H. 1983. Record dispersal by a wolf from Minnesota. *Journal of Mammalogy* 64:166–167.

Gaston, K. J. 1994. *Rarity.* London: Chapman & Hall.

Gilpin, M. E. 1987. Minimum viable populations: A restoration ecology perspective. In W. R. Jordan III, M. E. Gilpin, and J. D. Aber (eds.), *Restoration ecology: A synthetic approach to ecological research.* New York: Cambridge University Press.

Griffith, B., M. J. Scott, J. W. Carpenter, and C. Reed. 1989. Translocation as a species conservation tool: Status and strategy. *Science* 245:477–480.

Groom, M. J. 1998. Allee effects limit population viability for an annual plant. *American Naturalist* 151:487–496.

Gustaitis, R., and R. McGrath. 1991. Species wars. *California Coast and Ocean* 7(3):17–26.

Guzmán, H. M. 1991. Restoration of coral reefs in Pacific Costa Rica. *Conservation Biology* 5:189–195.

Hager, H. A., and K. D. McCoy. 1998. The implications of accepting untested hypotheses: A review of effects of purple loosestrife (*Lythrum salicaria*) in North America. *Biodiversity and Conservation* 7:1069–1079.

Handel, S. N., G. R. Robinson, W. F. J. Parsons, and J. H. Mattei. 1997. Restoration of woody plants to capped landfills: Root dynamics in an engineered soil. *Restoration Ecology* 5:178–186.

Hellmers, N. D. 1983. Hardwood forest restoration on Abraham Lincoln Farm, advice sought (Indiana). *Restoration and Management Notes* 1(4):23.

Holl, K. D., and J. Cairns Jr. 1994. Vegetational community development on reclaimed coal surface mines in Virginia. *Bulletin of the Torrey Botanical Club* 121:327–337.

Holt, A. 1994. Flora vs. fauna. *Vegetarian Times,* July 1994, pp. 75–78.

Horvitz, C. C. 1997. The impact of natural disturbances. In D. Simberloff, D. C. Schmitz, and T. C. Brown (eds.), *Strangers in paradise: Impact and management of nonindigenous species in Florida.* Washington, D.C.: Island Press.

Huenneke, L. F. 1991. Ecological implications of genetic variation in plant populations. In D. A. Falk and K. E. Holsinger (eds.), *Genetics and conservation of rare plants.* Oxford: Oxford University Press.

Jordan, W. R. III. 1983. Looking back: A pioneering restoration project turns fifty. *Restoration and Management Notes* 1(3):4–10.

Jordan, W. R. III, M. E. Gilpin, and J. D. Aber (eds.). 1987. *Restoration ecology: A synthetic approach to ecological research.* Cambridge: Cambridge University Press.

Kirkman, L. K., and R. Sharitz. 1994. Vegetation disturbance and maintenance of diversity in intermittently flooded Carolina bays in South Carolina. *Ecological Applications* 4:177–188.

Kline, V. M., and E. A. Howell. 1987. Prairies. In W. R. Jordan III, M. E. Gilpin, and J. D. Aber (eds.), *Restoration ecology.* Cambridge: Cambridge University Press.

Kunin, W. E. 1993. Sex and the single mustard: Population density and pollinator behavior effects on seed-set. *Ecology* 74:2145–2160.

Kunin, W. E., and K. J. Gaston (eds.). 1997. *The biology of rarity.* London: Chapman & Hall.

Lamont, B. B., P. G. L. Klinkhamer, and E. T. F. Witkowskodii. 1993. Population fragmentation may reduce fertility to zero in *Banksia goodii*—A demonstration of the Allee effect. *Oecologia* 94:446–450.

Lande, R. 1987. Extinction thresholds in demographic models of territorial populations. *Oecologia* 130:624–635.

———. 1993. Risks of population extinction from demographic and environmental stochasticity and random catastrophes. *American Naturalist* 142:911–928.

Loope, L. L., and P. G. Sanchez. 1988. Biological invasions of arid land nature reserves. *Biological Conservation* 44:95–118.

Louda, S. M., D. Kendall, J. Connor, and D. Simberloff. 1997. Ecological effects of an insect introduced for the biological control of weeds. *Science* 277:1088–1090.

Magnuson, J. J., H. A. Regier, W. J. Christie, and W. C. Sonzogni. 1980. To rehabilitate and restore Great Lakes ecosystems. In J. Cairns Jr. (ed.), *The recovery process in damaged ecosystems.* Ann Arbor: Ann Arbor Science Publishers.

Matson, P. A., and M. D. Hunter (eds.). 1992. Special feature: The relative contributions of top-down and bottom-up forces in population and community ecology. *Ecology* 73:723–765.

Mech, L. D. 1970. *The wolf.* Minneapolis: University of Minnesota Press.

Meda, D. 1990. Restoration in the Feliz Creek watershed, California. In J. J. Berger (ed.), *Environmental restoration: Science and strategies for restoring the earth.* Washington, D.C.: Island Press.

Mehrhoff, L. A. 1996. FOCUS: Reintroducing endangered Hawaiian plants. In D. A. Falk, C. I. Millar, and M. Olwell (eds.), *Restoring diversity: Strategies for reintroduction of endangered plants.* Washington, D.C.: Island Press.

Menges, E. S. 1992. Stochastic modelling of extinction in plant populations. In P. L. Fiedler and S. K. Jain (eds.), *Conservation biology: The theory and practice of nature conservation, preservation, and management.* New York: Chapman & Hall.

Middleton, B. A. 1995. Seed banks and species richness potential of coal slurry ponds reclaimed as wetlands. *Restoration Ecology* 3:311–318.

Miller, R. M. 1987. Mycorrhizae and succession. In W. R. Jordan III, M. E. Gilpin, and J. D. Aber (eds.), *Restoration ecology.* Cambridge: Cambridge University Press.

Miller, R. M., T. B. Moorman, and S. K. Schmidt. 1983. Interspecific plant association effects on vesicular-arbuscular mycorrhiza propagules during topsoil storage. *Journal of Applied Ecology* 22:259–266.

Mills, L. S., M. E. Soulé, and D. F. Doak. 1993. The keystone species concept in ecology and conservation. *BioScience* 43:219–224.

Minnich, R. A., M. G. Barbour, J. H. Burk, and R. F. Fernau. 1995. Sixty years of change in Californian conifer forests of the San Bernardino mountains. *Conservation Biology* 9:902–914.

Moreno, J. M., and W. C. Oechel (eds.). 1994. *The role of fire in Mediterranean-type ecosystems.* New York: Springer-Verlag.

Mount, J. F. 1995. *California rivers and streams: The conflict between fluvial processes and land use.* Berkeley: University of California Press.

Noss, R. F. 1992. The Wildlands Project: Land conservation strategy. *Wild Earth* (special issue) 1:10–25.

Owen-Smith, R. N. 1987. Pleistocene extinctions: The pivotal role of megaherbivores. *Paleobiology* 13:351–362.

———. 1988. *Megaherbivores: The influence of very large body size on ecology.* Cambridge: Cambridge University Press.

Paine, R. T. 1969. A note on trophic complexity and community stability. *American Naturalist* 103:91–93.

Paine, R. T., and S. A. Levin. 1981. Intertidal landscapes: Disturbance and the dynamics of pattern. *Ecological Monographs* 51:145–178.

Peterson, C. J., and S. T. A. Pickett. 1995. Forest reorganization: A case study in an old-growth forest catastrophic blowdown. *Ecology* 76:763–774.

Peterson, R. O., and R. E. Page. 1983. Wolf-moose fluctuation in Isle Royale National Park, Michigan, USA. *Annales Zoologici Fennica* 74:251–253.

Pickart, A. J. 1990. Dune restoration at Buhne Point, King Salmon, California. In J. J. Berger (ed.), *Environmental restoration: Science and strategies for restoring the earth*. Washington, D.C.: Island Press.

Platt, W. J., G. W. Evans, and S. L. Rathbun. 1988. The population dynamics of a long-lived conifer (*Pinus palustris*). *American Naturalist* 131:491–525.

Power, M. E., D. Tilman, J. Estes, B. A. Menge, W. J. Bond, L. S. Mills, G. Daily, J. C. Castilla, J. Lubchenco, and R. T. Paine. 1996. Challenges in the quest for keystones. *BioScience* 46:609–620.

Primm, S. A. 1996. A pragmatic approach to grizzly bear conservation. *Conservation Biology* 10:1026–1035.

Pruitt, B. C., and S. E. Gatewood. 1976. *Kissimmee River floodplain vegetation and cattle carrying capacity before and after canalization*. Special project to prevent the eutrophication of Lake Okeechobee. Tallahassee: Division of State Planning, Florida Department of Administration.

Pyne, S. J. 1982. *Fire in America: A cultural history of wildland and rural fire*. Princeton: Princeton University Press.

Pyne, S. J., P. L. Andrews, and R. D. Laven. 1996. *Introduction to wildland fire*. New York: Wiley.

Randall, J. M., R. R. Lewis III, and D. B. Jensen. 1997. Ecological restoration. In D. Simberloff, D. C. Schmitz, and T. C. Brown (eds.), *Strangers in paradise: Impact and management of nonindigenous species in Florida*. Washington, D.C.: Island Press.

Reeves, R. D., A. J. M. Baker, and R. R. Brooks. 1995. Abnormal accumulation of trace metals by plants. *Mining Environmental Management* 3 (September): 4–8.

Reinartz, J. A., and E. L. Warne. 1993. Development of vegetation in small created wetlands in southeastern Wisconsin. *Wetlands* 3:153–164.

Robbins, J. 1998. Engineers plan to send a river flowing back to nature. *New York Times,* 12 May, pp. B9 and B11.

Rodgers, T. 1997. Surgery breathes new life into Carlsbad lagoon. *San Diego Union-Tribune,* 2 July, p. B3.

Samuels, C. L., and J. A. Drake. 1997. Divergent perspectives on community convergence. *Trends in recent ecology and evolution* 12:427–432.

Samuels, C. L., and J. Lockwood. 1998. Restoration as an experiment in community assembly: A process-oriented approach. In R. E. Grese and D. Morrison (eds.), *Ecological restoration*. Washington, D.C.: Island Press.

Schmitz, D. C., and D. Simberloff. 1997. Biological invasions: A growing threat. *Issues in Science and Technology* 13(4):33–40.

Seliskar, D. M. 1995. Coastal dune restoration: A strategy for alleviating dieout of *Ammophila breviligulata*. *Restoration Ecology* 3:54–60.

Shrader-Frechette, K. S., and E. D. McCoy. 1993. *Method in ecology: Strategies for conservation.* Cambridge: Cambridge University Press.

Simberloff, D. 1986. The proximate causes of extinction. In D. M. Raup and D. Jablonski (eds.), *Patterns and processes in the history of life.* Berlin: Springer-Verlag.

———. 1990. Reconstructing the ambiguous: Can island ecosystems be restored? In D. R. Towns, C. H. Daugherty, and I. A. E. Atkinson (eds.), *Ecological restoration of New Zealand islands.* Wellington: New Zealand Department of Conservation.

———. 1994. Habitat fragmentation and population extinction of birds. *Ibis* 137:S105–S111.

———. 1996. Impacts of introduced species in the United States. *Consequences* 2(2):13–23.

———. 1998a. Flagships, umbrellas, and keystones: Is single-species management passé in the landscape era? *Biological Conservation* 83:247–257.

———. In press. Nonindigenous species: A global threat to biodiversity and stability. In P. Raven and T. Williams (eds.) *Nature and human society: The quest for a sustainable world.* Washington D.C. National Academy Press.

Simberloff, D., and P. Stiling. 1996. How risky is biological control? *Ecology* 77: 1965–1974.

Simberloff, D., D. C. Schmitz, and T. C. Brown (eds.). 1997. *Strangers in paradise: Impact and management of nonindigenous species in Florida.* Washington, D.C.: Island Press.

Smith, M. 1988. Reclamation and treatment of contaminated lands. In J. Cairns Jr. (ed.), *Rehabilitating damaged ecosystems.* Vol. 1. Boca Raton, Fla.: CRC Press.

Soulé, M. E. 1980. Thresholds for survival: Maintaining fitness and evolutionary potential. In M. E. Soulé and B. A. Wilcox (eds.), *Conservation biology: An evolutionary-ecological perspective.* Sunderland, Mass.: Sinauer.

———. 1990. The onslaught of alien species and other challenges in the coming decades. *Conservation Biology* 4:233–240.

———. 1994. Normative conflicts and obscurantism in the definitions of ecosystem management. In W. W. Covington and L. F. DeBano (eds.), *Sustainable ecological systems: Implementing an ecological approach to land management.* Fort Collins, Col.: USDA Forest Service.

Soulé, M. E., and R. F. Noss. 1998. Rewilding and biodiversity conservation: Complementary goals for continental conservation. *Wild Earth* 8(3):18–28.

Steele, B. B., and C. V. Grant. 1982. Topographic diversity and islands of natural vegetation: Aids in reestablishing bird and mammal communities on reclaimed mines. *Reclamation and Revegetation Research* 1:367–382.

Steenbergh, W. F., and C. H. Lowe. 1977. *Ecology of saguaro: III.* Scientific monograph series, no. 17. Washington, D.C.: National Park Service.

Tanji, E. 1993. Molokai hunters win removal of pig snares. *Honolulu Advertiser,* 20 March, p. 6.

Toth, L. A. 1993. The ecological basis of the Kissimmee River restoration plan. *Florida Scientist* 56:25–51.

———. 1995. Principles and guidelines for restoration of river/floodplain ecosystems—Kissimmee River, Florida. In J. Cairns Jr. (ed.), *Rehabilitating damaged ecosystems.* 2nd ed. Boca Raton, Fla.: CRC Press.

Towns, D. R., D. Simberloff, and I. A. E. Atkinson. 1997. Restoration of New Zealand islands: Redressing the effects of introduced species. *Pacific Conservation Biology* 3:99–124.

Trombulak, S. C. 1996. The restoration of old growth: Why and how. In M. B. Davis (ed.), *Eastern old-growth forests: Prospects for rediscovery and recovery.* Washington, D.C.: Island Press.

Tyler, C. M. 1996. Relative importance of factors contributing to postfire seedling establishment in maritime chaparral. *Ecology* 77:2182–2195.

U.S. Congress. Office of Technology Assessment. 1993. *Harmful non-indigenous species in the United States.* OTA-F-565. Washington, D.C.: Government Printing Office.

USDA Forest Service. 1989. *A guide for prescribed fire in southern forests.* Technical publication R8-TP 11. Atlanta: USDA Forest Service.

Veltman, C. J., S. Nee, and M. J. Crawley. 1996. Correlates of introduction success in exotic New Zealand birds. *American Naturalist* 147:542–557.

Walker, B. H. 1991. Biodiversity and functional redundancy. *Conservation Biology* 6:18–23.

Ward, S. C., J. M. Koch, and O. G. Nichols. 1990. Bauxite mine rehabilitation in the Darling Range, Western Australia. *Proceedings of the Ecological Society of Australia* 16:557–565.

Weaver, J. L., P. C. Paquet, and L. F. Ruggiero. 1996. Resilience and conservation of large carnivores in the Rocky Mountains. *Conservation Biology* 10:964–976.

Wegner, D. L. 1997. Let a river run through it. *Wild Rivers Review* (Berkeley) 12:4.

Westman, W. E. 1990. Park management of exotic plant species: Problems and issues. *Conservation Biology* 4:251–260.

Wharton, C. H., H. T. Odum, K. Ewel, M. Duever, A. Lugo, R. Boyt, J. Bartholomew, E. Debellevue, S. Brown, M. Brown, and L. Duever. 1977. *Forested wetlands of Florida—their management and use.* Gainesville: University of Florida Press.

White, P. S., and J. L. Walker. 1997. Approximating nature's variation: Selecting and using reference information in restoration ecology. *Restoration Ecology* 5:338–349.

Williams, T. 1997. Killer weeds. *Audubon* 99(2):24–31.

Williamson, M. 1996. *Biological invasions.* London: Chapman & Hall.

Woods, B. 1984. Ants disperse seed of herb species in a restored maple forest (Wisconsin). *Restoration and Management Notes* 2:29–30.

Wootton, J. T., M. S. Parker, and M. E. Power. 1996. Effects of disturbance on river food webs. *Science* 273:1558–1561.

Zedler, J. B. 1988. Salt marsh restoration: Lessons from California. In J. Cairns Jr. (ed.), *Rehabilitating damaged ecosystems.* Vol. 1. Boca Raton, Fla.: CRC Press.

Zimov, S. A., V. I. Chuprynin, A. P. Oreshko, F. S. Chapin III, J. F. Reynolds, and M. C. Chapin. 1996. Steppe-tundra transition: A herbivore-driven biome shift at the end of the Pleistocene. *American Naturalist* 146:765–794.

# 5 Core Areas: Where Nature Reigns

*Reed F. Noss, Eric Dinerstein, Barrie Gilbert,*
*Michael Gilpin, Brian J. Miller, John Terborgh,*
*and Steve Trombulak*

Experience on every continent has shown that only in strictly protected areas are the full fauna and flora of a region likely to persist for a long period of time. Hence the central component of a landscape design for conservation is the core area where human uses are greatly restricted and natural processes reign. We use the term "core area" to refer to areas where the conservation of biodiversity, ecological integrity, wilderness, or similar values takes precedence over other values and uses. In the context of continental reserve networks, core areas are designated largely to maintain existing natural qualities or to restore landscapes that have been degraded by human activities.

Some protected areas qualify as cores; others do not. Core areas might include national parks, designated wilderness areas on federal lands (the National Wilderness Preservation System), certain state or provincial parks and nature preserves (such as the wild forests of Adirondack Park in New York), reserves managed by The Nature Conservancy or other conservation groups, and, finally, other lands with an operational management plan that puts conservation of biodiversity or natural conditions as the highest priority (Noss and Cooperrider 1994). A distinguishing characteristic of core areas is limited human access—that is, low road density or, ideally, roadlessness. Many protected areas, including national parks in the United States and Canada (such as Banff in Alberta and Yosemite in California), vary in their degree of development and the extent to which biological conservation takes precedence over other values. Some portions of these parks are heavily developed with tourist facilities to the detriment of native wildlife. In these cases only the wilder, less accessible portions of the parks meet our definition of core areas.

Natural condition and limited access to humans are important attributes of core areas, but an area need not be pristine to qualify for protection as a core. In most of the eastern and midwestern United States, for example, a strict insistence on pristine conditions and roadlessness would disqualify virtually all areas from protection within the federal wilderness system. Candidate areas for wilderness designation—and core areas generally—need not be free of roads, evidence of past timber harvesting or mineral extraction, or other signs of human activity. Nevertheless, further road building, logging, and other damaging activities should be prohibited after designation.

In this chapter we review the purposes of core areas in conservation networks. We discuss methods of selecting core reserves and considerations for reserve design and management. From our review we conclude that core areas are essential for meaningful conservation and that rigorous methods should be used to select and design them. Focal species can be very useful in reserve selection and design, but they must be selected carefully using scientifically defensible criteria (B. Miller et al. 1988). As no single species can be expected to provide an umbrella effect for all other species and phenomena in the region of interest, usually several focal species must be selected. Furthermore, many cores will require restoration and other active management, at least in the short term, to attain their biological goals.

## Purposes of Core Areas

Core areas, when properly designed and managed, are places where nature can operate in its own way in its own time. They are also places where humans can encounter and understand nature in a relatively undefiled state. Lessons learned in studying these natural areas can be extended to management of other lands where resource exploitation and other intensive human uses take place.

### Biological, Scientific, and Other Values

Conservationists involved in reserve design must address several questions: What conservation goals are achieved in core areas that cannot be achieved by multiple-use management? Why must some core areas be large? What are the consequences—biological and social—of not having core areas? Although these questions should be addressed case by case, lessons learned through a number of case studies can inform new planning initiatives in their early stages.

Until recently most parks and other reserves were acquired for nonbiological reasons such as scenery, recreation potential, and lack of conflict with resource production objectives. Another important use of reserves, pointed out by Leopold (1941) but not appreciated by many land managers, is as a "base datum of normality" for a "science of land health" (see also Arcese and Sinclair 1997). In the context of ecosystem management (Agee and Johnson 1988; Grumbine 1994; Christensen et al. 1996), cores can serve as reference or control areas that can be compared to other areas receiving management treatments (for example, different kinds of logging). Although replication is not usually possible in such experiments and, too, most core areas are not entirely free of human influence, management experiments with no controls are dangerous. Too many ecosystem management experiments today are being conducted without controls. Similarly, reference areas have proved essential in setting standards for indices of biotic integrity for streams with varying degrees of degradation (Karr 1991). The larger and less accessible the core area (which for aquatic ecosystems should include entire watersheds whenever possible), the better it will function as a control that is rela tively free from human influences.

To provide adequate controls for ecosystem management experiments, an ideal reserve network would have landscape-scale reference areas for every major vegetation type or other ecosystem class native to the region. The present distribution of core areas is highly biased toward unproductive lands, however, and many types of ecosystems are poorly represented. For example, the GAP project for Idaho identified twenty-nine out of seventy-one vegetation types that were either not represented in protected areas or had less than 10,000 hectares represented (Caicco et al. 1995). Among forests, alpine types were well represented, but montane and low-elevation types were poorly represented. Similarly, in Sweden alpine landscapes were well represented but species-rich river landscapes were not (Nilsson and Götmark 1992). A preliminary gap analysis of abiotic zones (physical habitat classes) in the Klamath-Siskiyou region of northwestern California and southwestern Oregon found that 65 percent of the current protected area includes just three classes of habitats representing cool, high-elevation sites and areas with poor soils. Fully 47 percent of the coldest, poorest soil class is protected (some 26 percent of the total area in wilderness) compared to only 0.1 percent of the low-elevation, warm, best soil class (K. Vance-Borland, Conservation Biology Institute, unpublished data). For core areas to function effectively as controls, as critical habitats for key species, and as reservoirs of biodiversity generally (that is, as coarse filters), they should span entire environmental gradients and contain viable examples of all types of habitats (Noss 1987; Noss and Cooperrider 1994).

An essential function of core reserves is to serve the needs of species that are hypersensitive to human activities. Many of these species are large mammals—including both herbivores (for example, mountain sheep and goats, monkeys, tapirs) and carnivores (for example, wolverines). Large birds, such as eagles and both terrestrial and arboreal frugivores, are also highly sensitive in many cases, as are other species vulnerable to human persecution (such as snakes), hunting, or collecting for the pet or horticultural trade. For these species, indicators of human access—particularly road density—are often the most accurate predictors of habitat effectiveness (Lyon 1983; Thiel 1985; Noss and Cooperrider 1994: 54–57).

Recreational trails, even if only for foot travel, may also pose threats to some species (Knight and Gutzwiller 1995; S. G. Miller et al. 1998). As trail access increases, population density or reproductive success of sensitive species may decline. The sensitivity of large animals to human activity is often a learned response stemming from hunting pressure, legal or illegal. Paradoxically, some species considered classic wilderness animals, such as the grizzly bear, are often fairly tolerant of humans—the problem is that humans are intolerant of the bears (Mattson et al. 1996; Noss et al. 1996). In most regions where grizzly bears coexist with people, human-caused mortality is the leading category of bear deaths (Woodroffe and Ginsberg 1998). In one troubling case, grizzly bear mortality within a designated wilderness area exceeded mortality in the surrounding multiple-use lands because of heavy recreational hunting within the wilderness (Mace and Waller 1998). Larger wilderness areas with less trail access—or prohibition of hunting—would function better as true refugia for such animals.

Apart from maintaining populations of space-demanding and disturbance-sensitive species, large core areas with limited access will likely stand as the last biological fortresses against the invasion of exotic species (see Chapter 4). Exotic plants are less likely to invade undisturbed natural vegetation than vegetation disturbed by human activity. Whether this failure is generally attributable to competition from native species or to an inability to tolerate the natural process regime is not known. We do know that roads are often the avenues along which exotic plants (such as purple loosestrife), pathogenic fungi (such as Port Orford cedar root rot fungus), and certain animals (such as the gypsy moth) invade natural landscapes (Schowalter 1988; Noss and Cooperrider 1994; Forman 1995). Invasive exotics may disperse along disturbed roadsides or travel as seeds or spores in the tires, grills, or other parts of vehicles. In a car washing facility in Canberra, Australia, more than 18,500 seedlings representing almost 260 plant species were germinated from mud washed from vehicles (Wace 1977). The larger the core area and the lower the road density, the fewer invasions of exotics are expected.

We are not emphasizing here the economic values of protected areas because we do not believe they should be the main consideration in designating or managing cores. Relying on economic arguments to justify protection of core areas only encourages the practice of placing human interests above nature—a practice that created the biodiversity crisis in the first place. Nevertheless, we should recognize that protected areas often contribute greatly to local economies by providing pollination, protecting soil from erosion, regulating local climate, and protecting watersheds. For example, La Tigre National Park above Tegucigalpa, Honduras, produces about 40 percent of the water supply for the city—and the water is of high quality, requiring only limited processing (McNeely 1988). Tourism income is a salient economic benefit to many people living within or near protected areas.

Other values of core areas, not readily found in multiple-use or intensive-use zones, include providing opportunities for people to experience the wild. For many, wilderness is inspiring and renewing. Because it is often an uncomfortable and even dangerous place, wilderness can also be humbling. This humbling provides a positive benefit to the individual psyche and to civilization. Perhaps more important, people who feel a deep respect for nature—often bred in wild areas—are less likely to do violence against it.

## Criticism of Core Areas

Despite the widespread recognition among conservation biologists of the values of large, wild core areas, not everyone agrees. In recent years the trend in international conservation has been away from national parks and other core areas—which are often construed as imperialistic and elitist—toward projects called "sustainable development," "sustainable use," "ecodevelopment," or "integrated conservation-development projects" (ICDPs). This humanistic theme springs from the belief that protected areas in developing countries will survive only to the extent that they address human concerns (Western and Pearl 1989). A key assumption is that traditional communities use their resources in a sustainable manner (McNeely 1988). Most of these projects, however, have failed to maintain the integrity of core areas and appear to be leading to continued, often rapid, losses of biodiversity. (See Redford and Sanderson 1992; Robinson 1993; Kramer et al. 1997; Wells et al. 1999.)

In North America, moreover, there are people who dismiss the value of core areas and try to keep the amount of protected area to an absolute minimum. This tendency is not limited to the inane "wise use" movement and conservative politicians but has become the dominant attitude within some land managing agencies. Ecosystem management, as interpreted by federal and many state agencies in the United States, is distinctly utilitarian. In assuming that human stewardship of nature can be wise and fully cognizant of natural phenomena, ecosystem management demonstrates the "arrogance of humanism" that Ehrenfeld (1978) described. Most ecosystem management projects allocate little area to reserves. In the USDA Forest Service, for example, many scientists and managers have advocated what has been called a "landscape without lines" approach to ecosystem management—without any zoning and meaningful protected areas—and instead seek to manage for all uses across the entire landscape. The critical assumption in this approach is that managers have learned from past mistakes and are now capable of managing forests, rangelands, and other environments in an intelligent way compatible with biodiversity conservation. Ironically, the validity of this dubious claim can only be tested if, in fact, unmanipulated, landscape-scale core areas are available to serve as controls.

Two Forest Service scientists who have led the charge against protected areas recently explained their rationale as follows: "We think the reserve model is not flexible and often inefficient because it focuses too much on allocating the landscape to specific uses. . . . We should try to minimize balkanization of the landscape with permanent land allocations and standard prescriptions, and attempt to manage the landscape as an integrated whole" (Everett and Lehmkuhl 1997). This philosophy is exemplified by the interagency Interior Columbia Basin Ecosystem Management Plan released in 1997 and covering some 24 million hectares—including 70 percent of USDA Forest Service roadless areas in the Lower 48 states (J. Jontz, Western Ancient Forest Campaign, pers. comm. ). This egregious plan lacks any recommendations for new protected areas and instead proposes relatively minor changes in management practices. Curiously this project, as well as the earlier Northwest Forest Plan approved by President Clinton, was able to dispense with alternatives that included the establishment of meaningful protected areas by fashioning these alternatives to exclude all management (including restoration and prescribed burning). Hence the preservation alternatives in these planning processes were just straw men easily dismissed as ecologically unrealistic.

British Columbia provides an example of agency biologists questioning the applicability of the core-corridor-buffer model for grizzly bears. Hence the provincial management agency, with the belief that designating discrete cores would leave the intervening matrix without sufficient protection, has opted instead to manage the entire matrix for bears and humans. If the entire matrix were truly managed conservatively for protection of bears and other sensitive animals, this approach would indeed have merit. We doubt that this is feasible, however, and see no reason why protection of core areas and conservative management of the matrix should be seen as mutually exclusive. Both elements should be part of a comprehensive conservation strategy. Strictly protected core areas would provide a safety valve in the event that matrix management fails. As shown by Woodroffe and Ginsberg (1998) in their review of extinctions of large carnivores in protected areas that are too small, reserve border areas are often population sinks because they expose these animals to persecution by humans. Moreover, even if it works for bears, matrix management is not likely to provide the other values of core areas.

Attacks by resource managers on core areas come at a time when social scientists are dismissing normative ethics, wilderness, and even scientific facts as mere social constructs (Soulé and Lease 1994). This new brand of nihilistic relativism not only discards the values upon which conservation is based but also rejects the science that underpins conservation biology. Nevertheless, it is important to note that even those who recently have criticized the wilderness concept as ethnocentric, historically naive, and anachronistic, such as philosopher J. Baird Callicott (1994–95) and historian William Cronon (1995), have acknowledged the necessity of allocating lands to conservation purposes. Thus they implicitly denote a value for core areas that transcends philosophical debates over the extent to which our images of nature are social constructions.

Conservation strategies that lack meaningful core areas are naive, arrogant, and dangerous. Such approaches assume a level of ecological knowledge and understanding—and a level of generosity and goodwill among those who use and manage public lands—that are simply unfounded. Nearly all conservation biologists would reject ecosystem management plans that offer no meaningful protection to imperiled species and habitats. Yet many in government seem to have embraced the view that enlightened management can substitute for strict protection.

Hence scientists must educate the public and decision makers about the perils of the landscape-without-lines approach and point out the need for a greatly expanded system of core areas.

## Reserve Selection

We distinguish two phases in the establishment of core areas: reserve selection and reserve design. Selection has to do with identifying the key sites that should be included in a reserve network as core areas. Reserve design addresses the spatial configuration of the reserve network and includes consideration of lands, including corridors and buffers, that fall outside strict core areas (Noss and Cooperrider 1994; Scott and Csuti 1997).

Selecting lands to designate as core areas is not as straightforward as it might seem. In some regions the last undisturbed parcels are strikingly evident against a background of agricultural fields, suburbs, strip malls, or clearcuts. In other regions the natural remnants are larger and more options are available. In still other regions the landscape matrix remains as wildland, but with a broad gradient from highly degraded to nearly pristine habitat. In each of these scenarios the choice of areas to protect presents different challenges and opportunities. Selection also depends on whether the land is public or community land versus land owned by private individuals or corporations. Money for acquiring private land as core areas is limited (but see *Wild Earth,* vol. 8, no. 2), and political considerations limit the amount of public or private land that can be devoted to strict conservation.

In many cases land designated as a nature reserve has been dedicated for opportunistic or other nonscientific reasons—virtually guaranteeing that options to establish reserves will be exhausted before all elements of biodiversity are protected (Pressey et al. 1993). It is essential, then, that acquisitions target the lands of highest biological and ecological value: areas that are irreplaceable or otherwise stand to contribute most to conservation goals. New acquisitions should generally complement other acquisitions, both old and new, in terms of the specific elements protected so that the full range of biodiversity can be protected as efficiently as possible. Moreover, several occurrences of each species should be preserved because a single storm, epidemic, or other catastrophe could wipe out more populations.

Most parks, wilderness areas, and other reserves, then, have been acquired for nonbiological reasons. The application of biological goals and scientific principles to reserve selection began early in the 1920s and

1930s—most conspicuously within committees of the Ecological Society of America chaired by Victor Shelford. (See Shelford 1926, 1933; Kendeigh et al. 1950–1951; Noss and Cooperrider 1994. ) Science still has little leverage on the designation of many core areas. (Federal assessments of wilderness study areas in the United States focus largely on recreation visitor days.) Nevertheless, the influence of science is growing along with the increasing popularity of conservation biology. In recent decades, scientific approaches to core area selection have proceeded along three separate tracks: special elements, representation, or focal species. We review each of these tracks in turn.

## Special Elements

Special elements are species, places, and other entities considered to have high conservation value. Perhaps the best-known special elements are the "elements of diversity" ranked and tracked by the natural heritage programs (conservation data centers) established by The Nature Conservancy (Jenkins 1985, 1988; Noss 1987). The elements of greatest concern are those species and plant communities ranked as "critically imperiled globally" (G1) and "imperiled globally" (G2) based on their rarity and the severity of threats. "Element occurrences" are mapped locations of these elements. Conservation planners working at a regional or broader scale—the scales addressed in this book  generally look not for individual occurrences but for geographic clusters of occurrences. These clusters are often called hotspots, especially when they represent centers of endemism (Myers 1988). The assumption is that local land trusts and governments, or in some cases state agencies or The Nature Conservancy, can best protect the often small sites in which isolated endangered plants or other elements occur. Besides rare species and plant communities, there are other special elements that might be mapped in the process of selecting core areas: roadless areas and other relatively pristine sites; old-growth forests; unique geological sites; streams, lakes, and watersheds of high value for native fisheries or aquatic biodiversity; resource hotspots such as ice-free bays, artesian springs, and mineral licks; disturbance initiation or export areas (for example, peaks that receive lightning strikes and ridges that carry the resulting fires into new watersheds); sites recognized as sacred by indigenous peoples; sites inherently sensitive to development; and sites adjacent to existing protected areas.

This strategy has its weaknesses, however. In some regions hotspots may not stand out because a large number of endemic species are distributed throughout the region (as in much of Mexico). Moreover, the strategy assumes that everything we want to protect is still present and just waiting to be protected. In fact, some special elements—such as old-growth forests throughout most of North America—have been lost and will have to be restored. Nevertheless, remnant stands of old growth are of high value for providing blueprints for restoration and old-growth associated species to inoculate restoration zones with native species; hence they are worthy special elements.

## Representation

Protection of special elements, which often comprise the rare and unique in nature, does not assure that all species and habitats in a region will be adequately protected in core areas. In particular, species groups that are poorly known or inventoried may be missed. Hence the representation (or "coarse filter") strategy, which seeks to protect intact examples of each vegetation or habitat type in a region, can be considered complementary to special element protection. The assumption is that if we protect examples of all habitat types (complete environmental gradients), we will capture occurrences of a vast majority of species. (A further assumption, often implicit, is that gradients in species composition parallel gradients in physical habitat variables or that vegetation types, because they reflect environmental gradients, are surrogates for biodiversity.) Although the effectiveness of the coarse-filter approach has never been quantified (it would require complete species lists for all taxa in a region and would have to be repeated in different regions), it is reasonable. The gap analysis projects in the United States, Canada, and several other countries are examples of this approach (Scott et al. 1987, 1993).

The literature on representation, much of it originating in Australia, is large and dense and need not be reviewed here. Guidelines for assessing representation at an ecoregional scale have been presented by Noss (1996), and algorithms for efficiently selecting a complementary set of sites that represent all habitats in a region have been presented or reviewed by Pressey and Nicholls (1989), Bedward et al. (1992), Pressey et al. (1993, 1996), Church et al. (1996), Csuti et al. (1997), and many others. Questions have been raised, however, about the value of these approaches without consideration of the long-term viability of the selected sites. This concern has led leading researchers to develop new models that focus on retention as well as representation of biodiversity in reserves (Pressey 1998).

## Focal Species

The third means of selecting core areas is based on the needs of particular focal species. Focal species analysis answers many questions that cannot be addressed by considering special elements or representation. Whereas the locations of special elements and underrepresented habitats point to particular sites and landscapes that require protection, focal species analysis identifies additional high-value habitats and addresses two questions: How much area is needed? And in what configuration should habitat areas be designed? These questions are the bridge from reserve selection to reserve design.

Focal species serve many roles in conservation planning. A list of focal species that serves one conservation need may overlap only slightly with lists of species intended for other needs. Localized (endemic) species may be important special elements for identifying irreplaceable core areas in the selection process. Endemic species, however, may not be useful for other reasons such as low public interest and minuscule area and connectivity requirements.

People often assume that a particular focal species is an umbrella species—so that if its needs are met, all other species in the region will also persist. This is not necessarily so, however, though species with large area requirements are most likely to serve the umbrella function. There are other categories of focal species. Species that play pivotal roles in an ecosystem, disproportionate to their abundance, are known as keystone species. (See Chapter 3 for further discussion of focal species.) Some focal species are highly charismatic with the public and hence are useful as "flagships" stimulating broad support for conservation proposals. Some focal species are useful as ecological indicators because their responses to environmental change tell us something valuable about the integrity of the community at large. Although these species are sometimes of little use in reserve selection or design, they are valuable in monitoring and adaptive management (Lambeck 1997; Noss et al. 1997).

One of the first steps in using focal species for reserve selection is a clear description of their intended purposes. What species are being considered and why were they selected? What assumptions were made in the selection of those species? How will these focal species contribute to the general goals of the reserve network? What type and quality of data for each species are available? Because no single species can serve as a surrogate for the needs of all others, many biologists have recommended using a suite of species (Noss and Cooperrider 1994; Lambeck 1997; Noss et al. 1997).

Large carnivores often are chosen as the major focal species in regional and continental conservation strategies. In many cases this choice is well justified. Although large carnivores may or may not qualify as keystone species, apparently they often control populations of other species—such as ungulates and opportunistic mesopredators—that might otherwise have negative influences on the native biodiversity of an area. (See Chapter 3; Wilcove et al. 1986; Soulé et al. 1988; Terborgh 1988; Noss et al. 1996.) And carnivores, through their public appeal, can attract support for a conservation plan. In using terrestrial carnivores as umbrella species, it may be most practical to use the movements and spatial needs of females to estimate the number of animals a reserve could hold. Male carnivore movements are extensive, highly variable, and related mainly to social status, behavioral spacing mechanisms, and hormonal production (Ewer 1973; Powell 1979). Female carnivores are usually more valuable demographically and will raise their young in areas where critical resources are concentrated and easiest to obtain (Lindzey 1982; King 1989; Miller et al. 1996). Because they must satisfy elevated energy requirements with minimal time away from their young, they are restricted to optimal habitat. Therefore their home range sizes accurately represent the quality of that habitat (King 1989; Lindstedt et al. 1986). In highly fragmented habitat, however, considering only female needs can result in low mating success and high male mortality (Beier 1993). Moreover, it is usually male raptors that protect the breeding territories. The degree of emphasis placed on females depends on the natural history of the species chosen.

To what extent can a focal species serve as an umbrella for other species? This is a topic of considerable discussion in conservation biology. In calculating the umbrella function of the grizzly bear in Idaho, Noss et al. (1996) compared ambitious recovery zones (potential core and buffer areas) proposed by Shaffer (1992) to the distribution of vegetation types and vertebrate species in the Idaho Gap Analysis database (see Scott et al. 1993). Shaffer's proposed recovery zones covered 34 percent of the state of Idaho (compared to 8 percent in a U.S. Fish and Wildlife Service proposal). For two-thirds of the state's vegetation types, Schaffer's proposal would have protected at least 10 percent of their statewide ranges. In addition, it would have protected 71 percent of the mammal species, 67 percent of the birds, 61 percent of the amphibians, but only 27 percent of the reptiles native to Idaho. Thus large carnivores are likely to be better umbrellas for some communities, taxa, or species assemblages than for others. The proportion of species richness protected by considering individual species is likely to decline as one moves from a homogeneous landscape to a heterogeneous one with high beta-level diversity (see Chapter 2).

Until recently, most large carnivore populations ranged widely. In the Indian subcontinent tigers have declined but are still found in fifty-nine landscape units that contain a mixture of sanctuaries and government forests. Because their distribution overlaps with the most important areas for biodiversity in these landscapes, they make an ideal umbrella species for conservation planning (Dinerstein et al. 1999; Wikramanayake et al. 1998). But basing core areas on the present and truncated distribution of Amur tigers in the Russian Far East, which has been proposed, would exclude many hotspots of rare and endemic invertebrates and plants. A similar example in Africa is offered by Berger (1997), who shows that using the spatial needs of an existing herd of twenty-eight black rhinos did not assure healthy populations of six other herbivores. When Berger modeled the spatial needs of a hundred rhinos, the protection given to the six herbivores increased significantly. Kerr (1997) has found that basing reserves on the present distribution of complete carnivore communities did not offer good protection to several invertebrate taxa. Thus basing a reserve's location on present distribution of a species whose range has been reduced drastically by human activities, or using a species without a good understanding of the region's biogeography, can lead to poor decisions.

In general, then, we contend that an umbrella species approach is more suitable to questions of how much land is necessary in a reserve network and how that land should be configured (an issue of reserve design) than for where to place core areas—especially noteworthy considering the truncated distribution of many large carnivores today.

## Combining the Three Tracks

Although the three tracks of conservation planning are usually pursued in isolation, they are most useful when united into a single comprehensive approach (Noss 1996; Noss et al. 1999). (See Figure 5.1.) One promising approach to reconciling the three tracks is to develop a series of reserve design options for a region, each reflecting a different emphasis given to each of the three tracks. Conservation planners may decide that certain conservation criteria are not especially relevant to their region or congruent with their conservation goals. If rewilding (Soulé and Noss 1998) is the fundamental goal of a regional plan, for example, planners

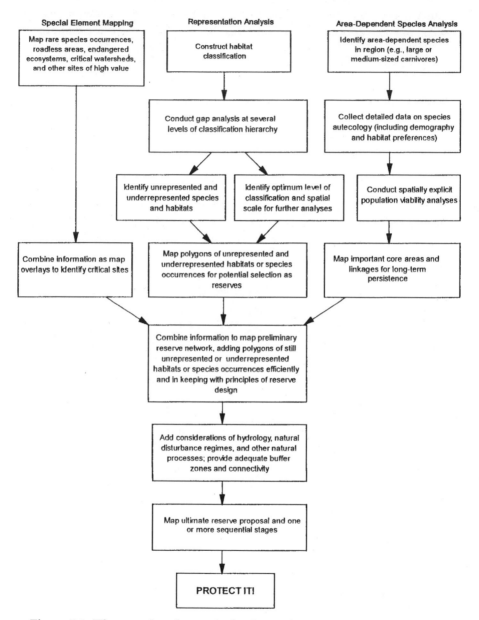

Figure 5.1. Three tracks of reserve selection and design, which rarely have been combined in practice. Adapted from Noss (1996).

may decide to dispense with representation analyses and mapping of rare species occurrences and concentrate instead on a restricted class of special elements—roadless areas, landscapes with high potential for road closures, habitat quality for prey species, connectivity—and a short list of focal species that are most area-demanding and sensitive to human activities (such as large mammalian predators). In regions lacking large, native carnivores but with high physical and biological heterogeneity (such as southwestern Australia) or in centers of endemism (such as many tropical and Mediterranean-climate regions), protection of endemic species and representation of plant communities may take precedence over rewilding. Hence the three-track approach is inherently flexible and adaptable to different goals and circumstances.

## Reserve Design

The design phase of reserve establishment follows on the heels of reserve selection and inevitably results in additional lands being recommended for protection. Ideally, core areas should be large enough to contain natural disturbance regimes and viable populations of top carnivores (at least over a span of decades). Where sufficiently large wildlands are not available, as in most of the Lower 48 states, smaller areas can serve as cores but will require greater investments in management as well as additional land allocated to buffer zones and linkages. These considerations highlight the need to think of core areas within the broad context of regional landscape design, which incorporates buffer areas (Chapter 7) and other human uses, as well as the critical issue of connectivity (Chapter 6). Nevertheless, a system of small cores is unlikely to attain ambitious conservation goals. Buffer areas, connectivity, and broad issues of reserve network design are addressed elsewhere in the book.

### Focal Species

Focal species, as we have seen, have a variety of potential uses in conservation planning. Large carnivores and other area-dependent species are particularly useful in designing reserve networks because of their demanding requirements for secure habitats and movement corridors. The needs of other appropriate focal species (see Lambeck 1997; Noss et al. 1997; B. Miller et al. 1998) should also be considered. Some biologists believe that the ideal approach is to conduct a rigorous population viability analysis (PVA) for each species based on real demographic and genetic data. To conserve all the biodiversity in a region, viability assessments of many focal species may be required. A multispecies model would add biological logic describing the interactions between pairs of the species, such as predation or interspecific competition for space. To protect species X, for example, it might be required that the overlap between the units of species X and Y be no more than 10 percent. Moreover, the cost of land acquisitions may be a consideration involving both economic and political factors. Using the techniques of discrete linear programming, a theoretically optimal reserve network can be designed. (Heuristic algorithms that offer "near-optimal" solutions with greater simplicity and practicality may be preferred in many cases; see Pressey et al. 1996; Csuti et al. 1997.) This analytic process does not result in a final map or proposal but rather a set of options and decision tools that planners can use in a stepwise fashion to move toward the final solution.

Ideal methodologies are sometimes impractical. Reserve design often cannot afford to wait until a PVA for each species of concern is completed. When the data, time, and funds to conduct PVAs are limited, other means of assessing the status and trends of focal species populations should be pursued (Noss et al. 1997; Carroll et al. 1998). Several approaches have been developed to make use of presence/absence information, which may be the only information available for many lesser-known species (Wiens 1989; Howe et al. 1991; Karieva et al. 1996; Hanski 1996; Hanski et al. 1996). These alternatives to intensive demographic studies and PVA are less data-hungry and can be used in many planning situations where timing is critical (Noss et al. 1997). An example is the spatially explicit incidence function model applied by Hanski et al. (1996) to a metapopulation of an endangered butterfly (the Glanville fritillary, *Melitaea cinxia*). The model does not require detailed demographic information but uses "snapshot" presence/absence data from a collection of habitat patches to estimate biologically important metapopulation parameters. The model predicted patch occupancy well and allowed for quantitative predictions about metapopulation dynamics.

Recognizing the uncertainty inherent in modeling, habitat should always be added to the reserve design to provide something more than the defined minimum required to provide for viable populations of each focal species. And it is important to recognize that the minimum viability unit for a large carnivore or other wide-ranging species is often considerably larger than the planning region—hence the need for interregional connectivity and coordination.

## Principles of Reserve Design

Reserve design, like all facets of conservation planning, is still as much art as science. The best algorithms for selecting core areas and determining their optimal distribution leave unanswered many practical questions of how to tie them all together into a design that offers a high probability of meeting conservation goals within a complex social and political framework. In practice, the conservation planner must rely on biological knowledge, empirical generalizations, political savvy, and intuition.

Various empirical generalizations—often called principles—of reserve design have emerged over the last two and a half decades. (See Terborgh 1974; Willis 1974; Diamond 1975; Wilson and Willis 1975; Diamond and May 1976.) Most of these principles deal with the requirements of area-sensitive focal species. These rules, and the controversy that has developed over them (Simberloff and Abele 1976), were to some extent reconciled by Soulé and Simberloff (1986), who agreed that reserves should be both large and multiple. Principles that have withstood the test of time and seem to apply to many cases have been proposed by Thomas et al. (1990) and Wilcove and Murphy (1991), with some modification by Noss (1992) and Noss et al. (1997). Among the principles that apply to reserve design are these:

- Species well distributed across their native range are less susceptible to extinction than species confined to small portions of their range.
- Large blocks of habitat, containing large populations, are better than small blocks with small populations.
- Blocks of habitat close together are better than blocks far apart.
- Habitat in contiguous blocks is better than fragmented habitat.
- Interconnected blocks of habitat are better than isolated blocks. (This principle addresses multiple blocks, often in different landscapes, whereas the preceding principle addresses the intactness of single blocks.)
- Blocks of habitat that are roadless or otherwise inaccessible to humans are better than roaded and accessible blocks.

All these rules have exceptions and their application to specific cases is not usually straightforward. They should not be applied uncritically. Proper interpretation of these principles can only be made by competent biologists familiar with the life forms and landscapes in question. Nevertheless, they provide a useful starting point for reserve design. In combination with site-specific and species-specific data and the judgment of experienced biologists, these principles can contribute to a design that has a high probability of meeting conservation goals.

## The International Context

We cannot overstate the importance of envisioning core areas within a broader geographic, biological, and social context. In developing countries, there are problems with poverty, lax protection of wildlife, and a large number of people who depend directly on the land for a meager survival. Few protected areas in developing countries have been effectively zoned to contain core areas as described in this chapter. Indeed, many reserves in Latin America have human settlements within their boundaries. Unless a part of a reserve is remote or virtually inaccessible, it is likely to be used by humans. Areas viewed by biologists as critical habitat for sensitive species, breeding areas for source populations of endangered species, or sites where key processes still occur are viewed by locals as a place to harvest trees and bush meat, cut grass, gather non-timber forest products, trap animals for trade, poach, or set fires to clear the forest for future hunting trips.

Recent efforts to link strict protection of protected areas with local use in many Asian settings have largely failed—delivering neither conservation nor development (Dinerstein et al. 1999; Wells et al. 1999). To conserve populations of endangered species that are habitat specialists or threatened by the illegal wildlife trade, there is no substitute for well-protected core areas. Although tigers are still distributed over large landscapes in India, for example, biologists suspect that tigers are only breeding in protected areas where there is still an adequate prey base; adjacent unprotected habitats are population sinks. Ecodevelopment projects that designate core areas without supporting strict protection measures inside them are building on quicksand (Dinerstein et al. 1999). A practical way to reduce pressure on core areas while increasing their effectiveness is to create economic incentives within adjacent buffer zones. To be effective, incentives must be commensurate with the spatial and temporal scale of the threats to core areas and directly related to those threats. If poaching is widespread across a core zone that is surrounded by villages, for example, incentives that go to a single village will do little to reduce poaching.

We suggest several approaches for enhancing protection of core areas drawn from ongoing projects in Nepal, some of which also apply to North America:

1.  Hire local people to protect buffer zone areas adjacent to the core area. This gives them a direct stake in conservation.
2.  Pass legislation that calls for the return of a large fraction of revenues to local development projects in the reserve buffer zone that are generated by park entry fees, concessions, and other enterprises that depend on a healthy core area.
3.  Promote alternatives to harvesting timber for firewood and building materials by managing areas in the buffer zone for the benefit of local people—and whenever possible allow them to take over management of these lands.
4.  Link reserve management efforts to rural development projects that introduce new environment-friendly technologies such as methane gas digestors and stall feeding of livestock.
5.  Fence and restore buffer zones and encourage community-based ecotourism efforts in buffer zones and corridors. This development can return significant sums to locals and divert tourism pressure from the core area.
6.  Link the concept of core areas to flood control, watershed protection, or other ecological services wherever appropriate and explain their value through public outreach programs.
7.  Link core areas with spiritual or religious symbolism and incorporate areas where taboos on hunting or entry prevail.
8.  Monitor the effects of incentives by using indicators of encroachment, poaching, and habitat degradation in core areas adjacent to buffer zones where incentive programs have been introduced.

Ultimately, in both industrialized and developing nations, core areas will succeed only when they enhance the economic value of adjacent buffer zones or other parts of the conservation network to local human populations or when the livelihoods of local people are directly tied to intact or well-managed core areas.

## Management of Core Areas

Management is anathema to conservationists who prefer a vision of nature untainted by human hands. It would be nice if there were a simple management plan that could reinstate the prehuman condition of biological evolution, but science is not able to provide such a prescription, at least not under the constraints imposed by our civilization.

## Toward Minimal Management

Short of complete, landscape-wide restoration, we espouse a policy of creating protected areas designed, whenever possible, to give free reign to abiotic and biotic processes critical to maintaining biodiversity. Our prescriptions include the reestablishment of natural fire regimes and river flows; the reintroduction of top carnivores, large ungulates, and other locally extirpated species; and, wherever possible, the elimination of alien species. These prescriptions will provide the closest attainable approximation to the original regime while recognizing that some processes, such as dispersal and migration, lie beyond current management capabilities. By letting natural processes prevail to the maximum practical extent, optimal conditions for the maintenance of biodiversity are provided at minimal cost in hands-on management. To minimize errors of commission, direct interventions are to be avoided wherever possible.

But achieving a regime where natural processes predominate will, in most of North America, take time and—paradoxically—human effort. Hence management is often necessary, but it should be driven by ecological sensibility. Management of core areas may be of two kinds: temporary (until an area is restored to the point where it can take care of itself) or perpetual. Because so little of North America is truly pristine and so many lands have been managed so poorly in the past, virtually all newly established core areas will need at least temporary management. Intensive, hands-on restoration will be required in many landscapes where natural processes and species composition have been severely altered by agriculture, logging, fire suppression, livestock grazing, elimination of large predators, invasions of exotic species, and other insults. A recent study of the causes of species imperilment in the United States determined that a large proportion of imperiled species is threatened by fire suppression and alien species (Wilcove et al. 1998). The authors conclude that "both types of threats must be addressed through active, 'hands-on' management . . . such as by pulling up alien plants and trapping alien animals or using prescribed fire to regenerate early successional habitats."

## Restoration

Restoration is a complex topic and, in some cases, the methods are still in the developmental stage. Chapter 4 discusses this topic in more detail. The first step is to stop whatever practice is degrading the natural ecosystem. Sometimes this may be enough to restore natural processes and conditions. In many cases, however, the wounds are deeper and restoration demands active intervention. Restoration techniques—including slope recontouring, soil preparation, planting, thinning, burning, and hydrological manipulation—are highly experimental and specific to certain plant communities and site conditions; detailed and systematic descriptions, however, are beyond our scope here. In some cases—in landscapes that were logged but not stocked with tree plantations, for example, and where sources for recolonization of native plants and animals remain—restoration might amount to simply closing roads (and ripping and revegetating roadbeds, as necessary), stepping back, and letting nature take its course. But the greater the disruption of natural processes by humans, the more human effort is required for restoration if we want native species to return to "normal" patterns of abundance and distribution within several human generations.

Reestablishing populations of keystone species (Chapters 3 and 4; Mills et al. 1992; Power et al. 1996) is a critical element in the restoration of natural processes in core areas. Even though management increasingly emphasizes an ecosystem perspective (Smith 1984; Scott et al. 1987; Noss and Cooperrider 1994; Pickett 1997), many individual species will require detailed study and specific attention (Noss et al. 1997). Managing keystone species provides an excellent opportunity to influence ecosystems through single-species management (Miller et al. 1994). Species that exert considerable influence on ecosystem processes include beavers, prairie dogs, bison, gopher tortoises, and many large carnivores.

Control of exotic plants and exotic or feral animals is often needed in core areas, especially small core areas with nearby roads or human settlements. Even in vast wildlands, however, feral ungulates in the form of sheep, goats, pigs, cattle, horses, and donkeys are established in many places in North America. Control campaigns and open hunting can reduce populations, but they rarely result in complete eradication. The optimal long-term solution is to allow native top predators to act as control agents. Even if exotic ungulates do invade core areas, predators may be able to keep their numbers in check so that their impact on native vegetation is kept to a minimum. But in arid rangeland regions—where large mammalian carnivores were generally not abundant prior to European settlement or, in some cases, appear to have been entirely absent—the direct removal of large, exotic herbivores (cattle, horses) that are not effectively controlled by coyotes may be necessary.

Ideally, core areas should contain the finest remaining examples of pristine native habitat. But in large areas of the continent, pristine natural habitats no longer exist or are, at best, too small and fragmented to sustain the full spectrum of diversity-maintaining processes. Lack of sufficient primary habitat should not be a pretext, however, for abandoning the effort to preserve native biodiversity. Core areas large enough to sustain critical processes should be established, wherever possible, because nature tends to restore itself with time. Top predators and large ungulates can be successfully reestablished in a disturbed landscape if conditions otherwise are amenable. Restoration of a natural process regime can be promoted by removing roads and dams, relaxing fire suppression, and curtailing hunting, grazing, logging, and mining. Whether the recovery of a natural process regime will eventually lead to the reduction or even elimination of exotic species remains a major untested hypothesis. Nevertheless we predict that reducing human disturbances, and directly removing exotics where they are found, are crucial for solving the exotics problem.

Management may be required in perpetuity for smaller, isolated core areas—say, those less than 400,000 hectares. Perpetual management may involve law enforcement, boundary maintenance, public education and visitor management, exotic species control, prescribed burning, and other habitat management to improve conditions for native species and communities, intensive population management (including translocations) of severely imperiled species, and more. The smaller the reserve, the more management per unit area is required (Bratton and White 1980; Noss 1983)—and the more it costs (Soulé 1984). This is one of the key reasons for establishing big reserves: they are usually more cost-effective to manage. And because less management and associated human presence are required, big reserves are more capable of sustaining conditions where visitors confront real wilderness rather than highly manipulated artifacts of nature.

## Monitoring and Adaptive Management

Monitoring is an essential component of management. The key to managing ecosystems is understanding biological relationships, processes, and change (Noss and Cooperrider 1994). Monitoring is also the only way to ascertain if the reserve design principles are correct. Moreover, monitoring and research aimed at addressing problems such as over-abundance of ungulates and decline of browsing-sensitive species (McShea et al. 1997) will be beneficial to other parks and reserves.

We should not regard reserve designs or management plans as final products. A plan should be an organized method of pursuing goals and opportunities and a way to improve our knowledge about what is needed to protect nature. If the process works, we will be able to respond swiftly and effectively to new problems as they arise. An adequate adaptive management framework, in the sense of Holling (1978), allows conservationists to solve problems and take advantage of opportunities. We emphasize that reserves of all kinds require law enforcement. Selective road removal and other controls of human access will ease the enforcement burden and reduce mortality for species hunted or persecuted by humans.

## The Need for Management

Conservation is biologically and socially complex and occurs in a context of uncertainty, change, and public scrutiny. Planning conducted in such an atmosphere is vulnerable to challenges that cut across institutional and geographical lines (Miller et al. 1996). Stakeholders often have disparate values, attitudes, and agendas. In such an arena, one group may think a problem is critical whereas a second faction shies away from it because it conflicts with their ideology. Consensus is almost impossible when groups have conflicting values. For this reason, a conservation plan is best created by a group of people with similar—though not necessarily identical—goals. Strong leadership and perseverance can make or break a plan.

The authors of this chapter, and indeed the entire book, share biocentric values. We agree that the native biodiversity of North America should be maintained and restored to the greatest extent possible. Hence we believe that conservation planning should occur on a regional to continental scale and that the present extent of protected area in North America is far too small to meet well-accepted conservation goals. We also recognize that given the historical evidence, rigorously protected core areas are an essential ingredient of a successful conservation strategy. Although some of us are species biologists and others are more interested in communities and ecosystems, we conclude that focal species are of great utility in the selection and design of core areas and their enveloping networks. These focal species, which must include a variety of taxa, should be selected and their habitat needs modeled on the basis of the best available information.

Perhaps one of our more controversial conclusions is that most core areas will require restoration—and, often, continued active management—if they are to function effectively. Although, in time, nature will restore itself, many species could be lost under a policy of benign neglect

(Soulé et al. 1979). Most conservation biologists appreciate the need for ecological management, but we have not communicated the message to nonscientists, including grassroots activists. Land management agencies such as the USDA Forest Service routinely exploit the aversion of conservationists for management. In both the president's Forest Plan for the Pacific Northwest and the Interior Columbia Basin Ecosystem Management Plan, for example, the alternative of calling for a substantial network of protected areas was a blatant straw man—a classic preservationist, hands-off alternative with no management of any kind allowed in reserves. Given the obvious need for ecological restoration in many of the forests and rangelands of these regions, evaluation panels easily rejected the preservation alternative, preferring the alternatives with active management and minimal protection. Motivated by similar values, anti-conservation journalists for *Science* and *Newsweek* magazines, in their articles on The Wildlands Project and the Yellowstone-to-Yukon initiatives, respectively, incorrectly labeled reserve proposals "hands off" or as human exclusion zones. The purpose of these caricatures was to discredit conservation biology and arouse distrust of conservation planning in the minds of the vocal segment of the public that favors both intensive resource production and recreation on public lands.

Conservation biologists must make one thing clear: where human activities have led to ecological degradation, it is a human responsibility to repair the damage. This is true both inside and outside core areas. Many of these reserves will need active management in perpetuity, as they are simply too small and isolated for fire, predator/prey relationships, and other ecological processes to operate within the range of variability that native species experienced over their evolutionary histories. Aggressive control or eradication programs are also necessary for many invasive, exotic species. It may be a bitter irony that human management is now needed to maintain some degree of naturalness in protected areas. This realization, however, does not detract from our ultimate vision of a wild, self-sustaining biosphere.

# References

Agee, J. K., and D. Johnson (eds.) 1988. *Ecosystem management for parks and wilderness.* Seattle: University of Washington Press.

Arcese, P., and A. R. E. Sinclair. 1997. The role of protected areas as ecological baselines. *Journal of Wildlife Management* 61:587–602.

Bedward, M., R. L. Pressey, and D. A. Keith. 1992. A new approach for selecting fully representative reserve networks: Addressing efficiency, reserve design, and land suitability with an iterative analysis. *Biological Conservation* 62:115–125.

Beier, P. 1993. Determination of minimum habitat areas and habitat corridors for cougars. *Conservation Biology* 7:94–108.

Berger, J. 1997. Population constraints associated with the use of black rhinos as an umbrella species for desert herbivores. *Conservation Biology* 11:69–78.

Bookbinder, M. P., E. Dinerstein, A. Rijal, H. Cauley, and A. Rajuria. Ecotourism's support of biodiversity conservation. *Conservation Biology* 12:1399–1404.

Bratton, S. P., and P. S. White. 1980. Rare plant management—after preservation what? *Rhodora* 82:49–75.

Caicco, S. L., J. M. Scott, B. Butterfield, and B. Csuti. 1995. A gap analysis of the management status of the vegetation of Idaho. *Conservation Biology* 9:498–511.

Callicott, J. B. 1994–95. A critique of and an alternative to the wilderness idea. *Wild Earth* 4(4):54–59.

Carroll, C., P. Paquet, R. Noss, and J. Strittholt. 1998. Modeling carnivore habitat in the Rocky Mountain region: A literature review and suggested strategy. Unpublished report. Corvallis, Ore.: Conservation Biology Institute.

Christensen, N. L., A. M. Bartuska, J. H. Brown, S. Carpenter, C. D'Antonio, R. Francis, J. F. Franklin, J. A. MacMahon, R. F. Noss, D. J. Parsons, C. H. Peterson, M. G. Turner, and R. G. Woodmansee. 1996. The report of the Ecological Society of America Committee on the Scientific Basis for Ecosystem Management. *Ecological Applications* 6:665–691.

Church, R. L., D. M. Stoms, and F. W. Davis. 1996. Reserve selection as a maximal covering location problem. *Biological Conservation* 76:105–112.

Cronon, W. (ed.). 1995. *Uncommon ground: Toward reinventing nature.* New York: Norton.

Csuti, B., S. Polasky, P. H. Williams, R. L. Pressey, J. D. Camm, M. Kershaw, A. R. Kiester, B. Downs, R. Hamilton, M. Huso, and K. Sahr. 1997. A comparison of reserve selection algorithms using data on terrestrial vertebrates in Oregon. *Biological Conservation* 80:83–97.

Diamond, J. M. 1975. The island dilemma: Lessons of modern biogeographic studies for the design of natural preserves. *Biological Conservation* 7:129–146.

Diamond, J. M., and R. M. May. 1976. Island biogeography and the design of natural reserves. In R. M. May (ed.), *Theoretical ecology: Principles and applications.* Philadelphia: Saunders.

Dinerstein, E., A. Rijal, M. Bookbinder, B. Kattel, and A. Rajuria. 1999. Tigers as neighbors: Efforts to promote local guardianship of endangered species in lowland Nepal. In J. Seidensticker, P. Jackson, and S. Christie (eds.), *Riding the tiger: Conserving tigers in a human-dominated landscape.* Cambridge: Cambridge University Press.

Ehrenfeld, D. 1978. *The arrogance of humanism.* New York: Oxford University Press.

Everett, R. L., and J. F. Lehmkuhl. 1997. A forum for presenting alternative viewpoints on the role of reserves in conservation biology? A reply to Noss (1996). *Wildlife Society Bulletin* 97:575–577.

Ewer, R. F. 1973. *The carnivores*. Ithaca: Cornell University Press.

Forman, R. T. T. 1995. *Land mosaics: The ecology of landscapes and regions*. Cambridge: Cambridge University Press.

Grumbine, R. E. 1994. What is ecosystem management? *Conservation Biology* 8:27–38.

Hanski, I. 1996. Metapopulation ecology. In O. E. Rhodes, R. K. Chesser, and M. H. Smith (eds.), *Population dynamics in ecological space and time*. Chicago: University of Chicago Press.

Hanski, I., A. Moilanen, T. Pakkala, and M. Kuussaari. 1996. The quantitative incidence function model and persistence of an endangered butterfly population. *Conservation Biology* 10:578–590.

Holling, C. S. 1978. *Adaptive environmental assessment and management*. London: Wiley.

Howe, R. W., G. J. Davis, and V. Mosca. 1991. The demographic significance of sink populations. *Biological Conservation* 57:239–255.

Jenkins, R. E. 1985. Information methods: Why the heritage programs work. *Nature Conservancy News* 35(6):21–23.

———. 1988. Information management for the conservation of biodiversity. In E. O. Wilson (ed.), *Biodiversity*. Washington: National Academy Press.

Karieva, P., D. Skelly, and M. Ruckleshaus. 1996. Reevaluating the use of models to predict the consequences of habitat loss and fragmentation. In S. T. A. Pickett, R. S. Otsfeld, M. Schachak, and G. E. Likens (eds.), *The ecological basis of conservation: Heterogeneity, ecosystems, and biodiversity*. New York: Chapman & Hall.

Karr, J. R. 1991. Biological integrity: A long-neglected aspect of water resource management. *Ecological Applications* 1:66–84.

Kendeigh, S. C., H. I. Baldwin, V. H. Cahalane, C. H. D. Clarke, C. Cottam, I. M. Cowan, P. Dansereau, J. H. Davis, F. W. Emerson, I. T. Haig, A. Hayden, C. L. Hayward, J. M. Linsdale, J. A. MacNab, and J. E. Potzger. 1950–51. Nature sanctuaries in the United States and Canada: A preliminary inventory. *Living Wilderness* 15(35):1–45.

Kerr, J. T. 1997. Species richness, endemism, and the choice of areas for conservation. *Conservation Biology* 11:1094–1100.

King, C. M. 1989. *The natural history of weasels and stoats*. Ithaca: Comstock Publishing Associates and Cornell University Press.

Knight, R. L., and K. Gutzwiller (eds.). 1995. *Wildlife and recreationists: Coexistence through research and management*. Washington, D.C.: Island Press.

Kramer, R., C. van Schaik, and J. Johnson. 1997. *Last stand: Protected areas and the defense of tropical biodiversity*. New York: Oxford University Press.

Lambeck, R. J. 1997. Focal species: A multi-species umbrella for nature conservation. *Conservation Biology* 11:849–856.

Leopold, A. 1941. Wilderness as a land laboratory. *Living Wilderness* 6(July):3.

Lindstedt, S. L., B. J. Miller, and S. W. Buskirk. 1986. Home range, time, and body size in mammals. *Ecology* 67:413–418.

Lindzey, F. G. 1982. Badger. In J. A. Chapman and G. A. Feldhamer (eds.), *Wild mammals of North America*. Balitmore: John Hopkins University Press.

Lyon, L. J. 1983. Road density models describing habitat effectiveness for elk. *Journal of Forestry* 81:592–595.

Mace, R. R., and J. S. Waller. 1998. Demography and population trend in grizzly bears in the Swan Mountains. *Conservation Biology* 12:1005–1016.

Mattson, D. J., S. Herrero, R. G. Wright, and C. M. Pease. 1996. Science and management of Rocky Mountain grizzly bears. *Conservation Biology* 10:1013–1025.

McNeely, J. A. 1988. *Economics and biological diversity: Developing and using economic incentives to conserve biological diversity.* Gland, Switzerland: World Conservation Union.

McShea, W. J., H. B. Underwood, and J. H. Rappole. 1997. *The science of overabundance: Deer ecology and population management.* Washington, D.C.: Smithsonian Institution Press.

Miller, B., G. Ceballos, and R. Reading. 1994. Prairie dogs, poison, and biotic diversity. *Conservation Biology* 8:677–681.

Miller, B., R. P. Reading, and S. Forrest. 1996. *Prairie night: Black-footed ferrets and the recovery of endangered species.* Washington, D.C.: Smithsonian Institution Press.

Miller, B., R. Reading, J. Strittholt, C. Carroll, R. Noss, M. E. Soulé, O. Sanchez, J. Terborgh, D. Brightsmith, T. Cheeseman, and D. Foreman. 1998. Focal species in the design of reserve networks. *Wild Earth* 8(4):81–92.

Miller, S. G., R. L. Knight, and C. K. Miller. 1998. Influence of recreational trails on breeding bird communities. *Ecological Applications* 8:162–169.

Mills, L. S., M. E. Soulé, and D. F. Doak. 1992. The history and current status of the keystone species concept. *BioScience* 43:219–224.

Myers, N. 1988. Threatened biotas: "Hot spots" in tropical forests. *Environmentalist* 8:187–208.

Nilsson, C., and F. Götmark. 1992. Protected areas in Sweden: Is natural variety adequately represented? *Conservation Biology* 6:232–242.

Noss, R. F. 1983. A regional landscape approach to maintain diversity. *BioScience* 33:700–706.

———. 1987. From plant communities to landscapes in conservation inventories: A look at The Nature Conservancy (USA). *Biological Conservation* 41:11–37.

———. 1992. The Wildlands Project: Land conservation strategy. *Wild Earth* (special issue)1:10–25.

———. 1996. Protected areas: How much is enough? In R. G. Wright (ed.), *National parks and protected areas.* Cambridge, Mass.: Blackwell.

Noss, R. F., and A. Y. Cooperrider. 1994. *Saving nature's legacy: Protecting and restoring biodiversity.* Washington D. C.: Island Press.

Noss, R. F., M. A. O'Connell, and D. D. Murphy. 1997. *The science of conservation planning: Habitat conservation under the Endangered Species Act.* Washington, D.C.: Island Press.

Noss, R. F., H. B. Quigley, M. G. Hornocker, T. Merrill, and P. C. Paquet. 1996. Conservation biology and carnivore conservation in the Rocky Mountains. *Conservation Biology* 10:949–963.

Noss, R. F., J. R. Strittholt, K. Vance-Borland, C. Carroll, and P. Frost. 1999. Science-based conservation planning in the Klamath-Siskiyou ecoregion. *Natural Areas Journal* 19.

Pickett, S. T. A. 1997. The ecological basis of conservation. In S. T. A. Pickett, R. S. Ostfeld, M. Shachak, and G. E. Likens (eds.), *The ecological basis of conservation: Heterogeneity, ecosystems, and biodiversity.* New York: Chapman & Hall.

Powell, R. A. 1979. Mustelid spacing patterns: Variations on a theme by Mustela. *Zeitschrift für Tierpsychologie* 50:153–165.

Power, M. E., D. Tilman, J. A. Estes, B. A. Menge, W. J. Bond, L. S. Mills, G. Daily, J. C. Castilla, J. Lubchenco, and R. T. Paine. 1996. Challenges in the quest for keystones. *BioScience* 46:609–620.

Pressey, R. L. 1998. If the world were a logical place . . . Principles for the effective location of scarce conservation resources. Paper presented at annual meeting of the Society for Conservation Biology, Sydney, Australia, 16 July.

Pressey, R. L., and A. O. Nicholls. 1989. Application of a numerical algorithm to the selection of reserves in semi-arid New South Wales. *Biological Conservation* 50:263–278.

Pressey, R. L., C. J. Humphries, C. R. Margules, R. I. Vane-Wright, and P. H. Williams. 1993. Beyond opportunism: Key principles for systematic reserve selection. *Trends in Ecology and Evolution* 8:124–128.

Pressey, R. L., H. P. Possingham, and C. R. Margules. 1996. Optimality in reserve selection algorithms: When does it matter and how much? *Biological Conservation* 76:259–267.

Redford, K. H., and S. E. Sanderson. 1992. The brief, barren marriage of biodiversity and sustainability? *Bulletin of the Ecological Society of America* 73:36–39.

Robinson, J. G. 1993. The limits to caring: Sustainable living and the loss of biodiversity. *Conservation Biology* 7:20–28.

Schowalter, T. D. 1988. Forest pest management: A synopsis. *Northwest Environmental Journal* 4:313–318.

Scott, J. M., and B. Csuti. 1997. Noah worked two jobs. *Conservation Biology* 11:1255–1257.

Scott, J. M., B. Csuti, J. D. Jacobi, and J. Estes. 1987. Species richness: A geographical approach to protecting future biodiversity. *BioScience* 37:782–788.

Scott, J. M., F. Davis, B. Csuti, R. Noss, B. Butterfield, C. Groves, J. Anderson, S. Caicco, F. D'Erchia, T. C. Edwards, J. Ulliman, and R. G. Wright. 1993. Gap analysis: A geographical approach to protection of biological diversity. *Wildlife Monographs* 123:1–41.

Shaffer, M. L. 1992. *Keeping the grizzly bear in the American West: A strategy for real recovery.* Washington, D.C.: Wilderness Society.

Shelford, V. E. (ed.). 1926. *Naturalist's guide to the Americas.* Baltimore: Williams & Wilkins.

———. 1933. Ecological Society of America: A nature sanctuary plan unanimously adopted by the society, 28 December, 1932. *Ecology* 14:240–245.

Simberloff, D., and L. G. Abele. 1976. Island biogeography theory and conservation practice. *Science* 191:285–286.

Smith, E. M. 1984. The Endangered Species Act and biological conservation. *Southern California Law Review* 57:361–413.

Soulé, M. E. 1984. Application of genetics and population biology: The what, where, and how of nature reserves. In *Conservation, science, and society.* Paris: UNESCO.

Soulé, M. E., and G. Lease (eds.). 1994. *Reinventing nature? Responses to postmodern deconstruction.* Washington, D.C.: Island Press.

Soulé, M. E., and R. F. Noss. 1998. Rewilding and biodiversity: Complementary goals for continental conservation. *Wild Earth* 8(3):18–28.

Soulé, M. E., and D. Simberloff. 1986. What do genetics and ecology tell us about the design of nature reserves? *Biological Conservation* 35:19–40.

Soulé, M. E., B. Wilcox, and C. Holtby. 1979. Benign neglect: A model of faunal collapse in the game reserves of East Africa. *Biological Conservation* 15:259–272.

Soulé, M. E., D. T. Bolger, A. C. Alberts, J. Wright, M. Sorice, and S. Hill. 1988. Reconstructed dynamics of rapid extinctions of chaparral-requiring birds in urban habitat islands. *Conservation Biology* 2:75–92.

Stanley, T. R. 1995. Ecosystem management and the arrogance of humanism. *Conservation Biology* 9:255–262.

Terborgh, J. 1974. Preservation of natural diversity: The problem of extinction prone species. *BioScience* 24:715–722.

———. 1988. The big things that run the world: A sequel to E. O. Wilson. *Conservation Biology* 2:402–403.

Thiel, R. P. 1985. Relationship between road densities and wolf habitat suitability in Wisconsin. *American Midland Naturalist* 113:404–407.

Thomas, J. W., E. D. Forsman, J. B. Lint, E. C. Meslow, B. R. Noon, and J. Verner. 1990. A conservation strategy for the northern spotted owl. Portland, Ore.: USDA Forest Service, USDI Bureau of Land Management, USDI Fish and Wildlife Service and USDI National Park Service.

Wace, N. 1977. Assessment of dispersal of plant species—the car borne flora in Canberra. *Proceedings of the Ecological Society of Australia* 10:166–186.

Wells, M., S. Guggenheim, A. Khan, W. Wardojo, and P. Jepson. 1999. *Investing in biodiversity: A review of Indonesia's integrated conservation and development projects.* Washington, D.C.: World Bank.

Western, D., and M. C. Pearl (eds.), 1989. *Conservation for the twenty-first century.* New York: Oxford University Press.

Wiens, J. A. 1989. *The ecology of bird communities.* Vol. II: *Processes and variations.* New York: Cambridge University Press.

Wikramanayake, E. D., E. Dinerstein, J. G. Robinson, U. Karanth, A. Rabinowitz, D. Olson, T. Mathew, P. Hedao, M. Conner, G. Hemley, and D. Bolze. 1998. An ecology-based method for defining priorities for large mammal conservation: The tiger as a case study. *Conservation Biology* 12:865–878.

Wilcove, D. S., and D. D. Murphy. 1991. The spotted owl controversy and conservation biology. *Conservation Biology* 5:261–262.

Wilcove, D. S., C. H. McLellan, and A. P. Dobson. 1986. Habitat fragmentation in the temperate zone. In M. E. Soulé (ed.), *Conservation biology: The science of scarcity and diversity.* Sunderland, Mass.: Sinauer.

Wilcove, D. S., D. Rothstein, J. Dubow, A. Philips, and E. Losos. 1998. Quantifying threats to imperiled species in the United States. *BioScience* 48:607–615.

Willis, E. O. 1974. Populations and local extinctions of birds on Barro Colorado Island, Panama. *Ecological Monographs* 44:153–169.

Wilson, E. O., and E. O. Willis. 1975. Applied biogeography. In M. L. Cody and J. M. Diamond (eds.), *Ecology and evolution of communities*. Cambridge, Mass.: Belknap Press.

Woodroffe, R., and J. R. Ginsberg. 1998. Edge effects and the extinction of populations inside protected areas. *Science* 280:2126–2128.

# 6 Corridors: Reconnecting Fragmented Landscapes

*Andy Dobson, Katherine Ralls, Mercedes Foster,*
*Michael E. Soulé, Daniel Simberloff, Dan Doak,*
*James A. Estes, L. Scott Mills, David Mattson,*
*Rodolfo Dirzo, Héctor Arita, Sadie Ryan, Elliott A.*
*Norse, Reed F. Noss, and David Johns*

If one stands in the middle of a national park or wilderness area, such as Yellowstone National Park or Lake Nakuru in Kenya, it is still possible to believe that we have done an excellent job of conserving nature. Yet if one hikes to the top of the mountains that surround these parks, then one sees just how abruptly the park ends. The western boundary of Yellowstone is sharply delineated by clear-cutting of the forests outside the park. The western boundary of Lake Nakuru is marked by a patchwork of small farms and an increasing number of huts and other dwellings that form the advancing front of a nearby town.

Except for some boreal regions, this pattern is becoming ubiquitous—a sign of the major threat to biodiversity. A century or two ago in North America and in most of the tropics as well, human settlements were the islands and nature was the sea. Now it is the reverse. We see it in the Serengeti in Tanzania, Bukit Barisan in Sumatra, Minsmere in eastern England, or Monteverde in Costa Rica. Human activities around parks and wilderness areas are rapidly leading to their isolation as "islands of natural habitat in a sea of human development" (Janzen 1986; 1989). This fragmentation of nature disrupts the natural movements of animals, the seeds, spores, and pollen of plants, not to mention nutrient and energy flows within and between different sections of the landscape.

Simultaneously, activities outside of protected areas insidiously intrude on the functioning ecosystems within.

The interactions between protected areas and the developed areas outside them are often subtle and ill defined. Increased isolation is likely to lead to changes in the flow of nutrients and pollutants into and out of the protected area. Habitat conversion in the surrounding matrix eases the invasion of alien species that compete with, or prey upon, the species that the reserve was designed to protect. Isolation of wildlands can also disrupt migration corridors that are used by seasonal residents of the park. Soon species begin to vanish.

The problem is often referred to as habitat fragmentation—one of the most challenging issues facing conservationists. Fragmentation occurs at various scales: at a local or regional scale, the different habitats required by a species may be separated by dwellings, shopping malls, clear cuts, cities, and extensive agricultural development; at another extreme, areas used by migratory species for breeding and wintering grounds are often separated by enormous expanses of inappropriate habitat—either as a result of habitat conversion or, rarely, by natural forces such as lava flows.

It is widely recognized that habitat loss and habitat fragmentation are the major proximal threats to biological diversity (Heywood 1995; Laurance and Bierregaard 1997; Wilcove et al. 1986). The demographic consequences of isolation versus connectivity have also been considered, at least since the work of Nicholson and Bailey (1935) and Andrewartha and Birch (1954). (See reviews by Hanski and Gilpin 1991; Hastings and Harrison 1994; Hanski and Simberloff 1997.) Habitat fragmentation, in particular, brings about the isolation of populations, which in turn may lead to local population and species extinction and, in general, to reductions in biological diversity. Loss of keystone species leads to the disruption of ecosystem functions such as pollination, seed dispersal, and nutrient cycling and to population explosions in herbivores such as deer and rodents and small carnivores such as raccoons. And these in turn cause further extinctions (Chapter 3). Ultimately the twin processes of habitat loss and fragmentation lead to the replacement of original ecological communities by an impoverished, partly cosmopolitan, biota.

In some instances it is possible to maintain connectivity between the fragments by protecting natural corridors or stepping stones of appropriate habitat. In other cases, natural habitat and stopover points are gone; in this event, enhancement of converted habitats through ecological restoration becomes crucial (Chapter 4). The restoration of connectivity must occur at many scales. Most attention to date has focused on local and regional connectivity as a way to redress the various forces threatening small populations. Amphibians, for example, must be able to move safely across a country lane during their annual migration to a breeding pond. Yet many regional and interregional corridors, including those envisioned in The Wildlands Project, have different goals—such as accommodating the need for grizzly bears to disperse safely between the Canadian Rockies and the northern Rockies of the United States. In the face of global climate change and other major environmental shifts, a substantial system of landscape connections is a major prerequisite for ensuring species persistence (Hunter et al. 1988; Peters 1988).

In this chapter we emphasize the larger, regional, scale of connectivity. Our overall goal is to determine ways to connect the natural parts of a landscape that ensure the continuing function of the ecosystems within it. An important step is to review the empirical and theoretical evidence for the benefits of connectivity.

## What Is a Corridor?

Before discussing the evidence for the benefits of corridors, we should define the terms "corridor," "connectivity," and related concepts. Corridors are formally defined in landscape ecology as linear habitats that differ from the extensive matrix in which they are embedded (Forman and Godron 1986). This definition is vague, however, and has multiple meanings. Even when restricted to mean a landscape feature that fosters connectivity, "corridor" can connote different landscape features. Often it means simply a pathway by which animals move (Bennett 1990), but it can also constitute a habitat in which animals can feed and breed as well as move through. Indeed for a plant population to achieve connectivity by this means, a corridor would have to be not just a movement corridor but suitable habitat. The term "connector" seems to be used synonymously with "movement corridor." So do "landscape linkage" (Hudson 1991) and "link" (Gustafsson and Hansson 1997), apparently chosen for the sake of alliteration rather than to avoid controversy.

Some meanings of corridor are independent of population connectivity, though these different functions have often been conflated in the connectivity literature. (See the references in Simberloff et al. 1992.) For example, natural habitats that happen to occur in linear configurations, such as riparian communities, are occasionally called "corridors," as are artificial habitats such as highways, power lines, and railroad rights-of-way. Greenbelts and buffers, essentially aesthetic amenities, are sometimes called "corridors," too, perhaps in response to the increasing popularity of the term in conservation circles. Large landbridges such as the Isthmus of Panama have been termed "corridors." So too have underpasses and tunnels designed to prevent roadkill, although this usage has waned. Even arrays of discrete refuges for migratory waterfowl are sometimes called "corridors" (Date et al. 1991). Finally, it should be noted that the greenways movement has historically had an entirely different concern—recreation and aesthetics—and originated in the nineteenth century with Frederick Law Olmsted's "park and parkway" idea (Little 1990). The burgeoning interest in greenways stems from their amenity value, not their utility for wildlife—as in "The Cross Florida Greenway: A Mid-State Connector for Florida's Statewide System of Greenways" (Florida Greenways Commission 1994).

To some extent the confusion about the meaning of "corridor" is created by the everyday use of the word—a passage between two places. The image that springs to mind is that of a long, narrow path, like a hallway, connecting two patches of habitat used by animals. There is evidence, for example, that mountain lions can use fairly narrow habitat remnants that are not in fact good lion habitat (Beier 1996). But for most animals the popular image evoked by the word—a simple passage or transit way—is misleading. Many species, to successfully move between patches of suitable habitat, need linkages that are themselves good or at least marginal habitat and will accommodate normal patterns of movement without bringing the animal into contact with sources of high mortality, such as roads. Thus a grizzly bear may need a wide corridor— and wider if longer—with sufficient food, water, cover and very low density of roads. The word "corridor" often evokes a misleading image among scientists and conservationists alike. Terms such as connectivity and linkage might better represent the goal of maintaining ecosystem viability. Connectivity and linkage have fewer conceptual associations and allow us to explain what we mean without having to overcome deep-seated linguistic habits. Throughout this chapter the term "corridor" denotes large, regional connections that are meant to facilitate animal movements and other essential flows between different sections of the landscape.

Although attention has focused on the function of corridors as pathways for individual movement and interchange, they may also function as wildlife habitat if they provide the resources needed for some, if not all, of an organism's life-cycle activities (Downes et al. 1997; Rosenberg et al. 1997). Such corridors may function as "source" patches for some species, providing surplus individuals to patches of lower quality. In this case, the distinctions between core and corridor begin to disappear—particularly if survivorship, reproduction, and recruitment rates are high. Downes et al. (1997) found that total density of mammals was higher in corridors than in eucalyptus forests, indicating that, in some cases, corridors may indeed provide important wildlife habitat. It should be obvious, too, that a given corridor will almost always be habitat for many small species, though it may not serve as habitat for larger animals. Note that the opposite of "source" is "sink": a patch where the mortality rate is relatively high, potentially draining individuals from nearby areas. Sometimes population sinks—such as road edges that facilitate poaching—are juxtaposed next to productive habitat for the species. Planners and managers should avoid such situations.

## The Benefits of Landscape Connectivity

The retention or restoration of connections between patches has been considered a practical tool that may ameliorate the effects of habitat fragmentation on wildlife (Frankel and Soulé 1981; Hudson 1991; Saunders and Hobbs 1991). Spatial connectors may provide habitat for the maintenance of species within an altered landscape and may also function as a pathway for the movement and exchange of individuals among otherwise isolated habitat remnants. The mechanisms whereby such movement may promote local persistence can be demographic and genetic. The former may include a decline in the variability of birth and death rates and an increase in the rates of colonization and recolonization of remaining habitat patches. The demographic "rescue effect" is an example of the benefits of immigration (Brown and Kodric-Brown 1977). The advantages of genetic interchange may include a decline in inbreeding depression and an increase in potentially adaptive genetic variance.

Corridors and connections in the landscape have two major functions at the species level. First, they permit regular daily or seasonal movements—thereby helping to ensure that different subsections of larger populations have access to all the resources they require while also maintaining the potential for all individuals in the population to interbreed successfully. The regular movements range from the diurnal movement of snakes, birds, and many other creatures, between feeding and nesting or brooding sites, to the annual migrations of large ungulates. Second, connectivity facilitates the dispersal of animals from their place of birth to their adult home range where they breed. At the regional or landscape level, this latter function is usually the most important. It justifies, for example, connectivity between mountain ranges in the western United States.

## Riparian Corridors

Riparian zones appear to have considerable potential as "natural corridors" at the regional or landscape scale (Naiman et al. 1993). In upland zones they form natural connections between highland areas and the surrounding low-lying lands. In arid areas many species are drawn to the resources these areas maintain. In both cases, riparian corridors are of vital importance in moving water and dissolved nutrients between habitats. Studies of riparian plant communities in the Peruvian Amazon, Scandinavia, France, and the northeastern United States all report high levels of vascular plant diversity—for example, 20 percent of 4000 to 5000 Amazonian tree species are found in the flooded forest (Junk 1989). The high diversity in these areas is largely a result of the interaction between small-scale variations in topography and the disturbance regimes caused by the intensity and frequency of floods. These processes collectively create a nonequilibrium mosaic of habitats that allow a large variety of species to coexist.

Much of the landscape throughout the temperate and tropical world demonstrates the dire consequences of destroying riparian corridors by diverting river water for agricultural and industrial uses. But pressure on surviving riparian habitats will only increase: in many of the world's arid regions, more than 50 percent of the human population lives within 1 kilometer of a riparian corridor and uses it for most of their daily needs (Turner et al. 1990). The current enthusiasm for restoring riparian corridors, however, often in areas where they never existed, has created its own class of problems. The westward migration of cowbirds and other nonnative bird species into the eastern Rockies, for example, has been largely facilitated by the current zeal for riparian corridor restoration in the western plains. This suggests a greater emphasis should be placed on protecting threatened riparian habitats that currently act as corridors, rather than creating artificial riparian connections that might facilitate the movement of undesirable species.

## A Tropical Example

An example of the relevance of propinquity and connectivity for insects and plants is provided by recent studies of tropical fig species and their pollinators. Fig trees (*Ficus* spp.) and their specialized wasp pollinators (Agaonidae, Agaoninae) constitute pairs of obligate mutualists whereby the pollination and reproductive success of trees depends on the local movements of their pollinators. In turn, the survival of the wasps depends on the year-round availability of sexually mature fig trees. The largely tropical genus *Ficus* comprises many species, and each species has a specific agaonid wasp pollinator (Janzen 1979; Ramirez 1970; Nason et al. 1998).

The tight interdependence of these organisms makes them particularly vulnerable to local extinction in small, isolated patches of forest. Fig flowers are enclosed in a complex inflorescence or syconium (the fig "fruit") that attracts the females of the specific pollinator during the tree's female phase. Female wasps pollinate the flowers inside the fig and lay their eggs in the ovaries of some. Females wasps die inside the syconium after oviposition, and weeks later the seeds and larvae complete their development. The fig then enters a male phase coincident with the emergence of adult wasps, which mate within the fig. After mating, male wasps die; females collect pollen and leave the fig in search of another fig in which to pollinate flowers and lay their eggs.

Flowering is synchronous within an individual fig tree, so each generation of wasps must locate a different host tree. The cycle can be maintained only if receptive trees (with flowers) are available year-round. Given that tropical figs typically occur at low densities, the survival of the system depends on the existence of a habitat patch large enough to contain the critical minimum number of trees compatible with the local movement capabilities of the wasps (modeled under different biotic scenarios by Anstett et al. 1997 and Nason et al. 1998). Mature fig "fruits" are an important food resource for a large number of animals, which are in turn essential dispersal agents of other plant species. Thus figs are considered to be a tropical keystone resource (Terborgh 1986). For systems like this one, core areas (or remaining fragments) must be large enough to accommodate both the animals' local movements and sufficient numbers of codependent plants to meet the animals' needs—or there must be corridors to link smaller patches and thereby increase reserve size sufficiently to meet those needs.

## Scale and Design

Perhaps the major consideration for reserve designers in establishing connections for dispersal is that different species have different requirements. Nocturnal species such as bats and small carnivores will have little difficulty crossing gaps that are several hundred meters wide—although in the case of rodents, the presence of large gaps could increase risk from predation by owls and other nocturnal predators. Diurnal mammals and reptiles, however, will find ways to avoid these gaps. Many birds in tropical forests and chaparral habitat are reluctant to cross gaps. Hence a corridor for one species may present a barrier for another, and these differences in behavior must be carefully examined. While mammals and birds will actively disperse, most invertebrates and plants disperse passively and are either wind-borne or carried by animals that act as seed dispersers.

Wind-borne forest species increase their dispersal distance considerably in a fragmented and open landscape; by contrast, animal-borne species may indirectly depend on corridors for their seed dispersers.

The dimension of corridors is a perennial and complex subject. Although there is considerable debate over their optimal width, plainly the ratio of width to length should increase as we move from the hedgerow to the regional scale. At the largest scale, The Wildlands Project recommends that connections should be at least three times as wide as the distance to which edge effects are likely to extend. Harrison (1994) has suggested that the home ranges of large species should be used to set the minimum width of connections between patches of habitat. A comprehensive survey of home range size suggests that the home range requirements of carnivores, primates, and ungulates scales allometrically with body size (Peters 1983). Thus large mammals—and particularly large carnivores and birds of prey— require large home ranges (Figure 6.1). Further examination of these underlying allometric relationships may yield important quantitative insights into the minimum width of landscape connections in different regions.

But these empirical estimates of minimum width for corridors are

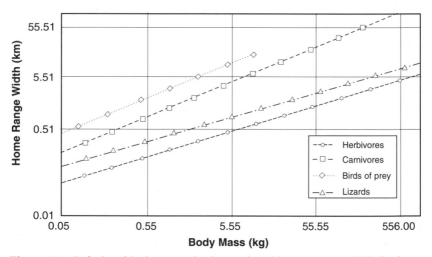

Figure 6.1. Relationship between body mass and home range width for herbivorous mammals, birds of prey, and lizards. The data for lizards were originally plotted by Turner et al. (1969); for birds by Schoener (1968); and for mammals by Harestad and Bunnell (1979). See Peters (1983) for a full discussion of the relationship between body size and home range.

tempered by the sobering recent work of Woodroffe and Ginsberg (1998) indicating that poaching is the greatest threat to the viability of large carnivores in many of the world's national parks. Their analysis implies that if corridors for carnivores are less than one home range in width, they will essentially lead carnivores (and edible ungulates) into the snares and traps of poachers. This finding also applies to buffer zones (Chapter 7) where carnivores—and other potential keystone species—are unlikely to persist because of human access. This suggests that truly viable landscape connections for edible or carnivorous vertebrates will have to be very large indeed, at least where poaching is virtually uncontrollable.

## Categories of Movement

Plans to reconnect the landscape must consider the full range of different species and the ecosystem functions they mediate. The spatial requirements of animals are dictated by their movement patterns, which are largely a function of their body size, ecology, and behavior. Movement patterns fall into four general categories: local movements; dispersal; nomadism; and seasonal migrations. We can examine these patterns from two spatial perspectives—those of human observers and those of the species in question.

Human beings develop systems of measurement based on the size of their appendages and a walking pace. Hence we classify movements on a continuum from those that span habitat patches (meters) to those that span regions (hundreds of kilometers). For example, the local or hedgerow scale (Noss's fencerow scale) connects small, close patches of habitat to facilitate the movement of invertebrates, plant seeds, and small vertebrates such as mice, voles, and chipmunks. Corridors at this scale are entirely edge habitat and encourage diversity at a very local level. Yet they can significantly increase and maintain the diversity of a region. A rough rule of thumb equates the diversity of tree and shrub species in an English hedgerow to its age: Hooper's rule says that the age of a hedge in centuries is approximately equal to the number of species located in a 27.5-meter section (Pollard et al. 1974; Moore et al. 1967). In England the removal of hedgerows surrounding fields has contributed significantly to the decline of bird and insect diversity in the last fifty years.

Hedgerows, windbreaks, and living fence posts, as well as ornamental plantings in parks and around buildings, can serve the purpose of enhancing movement between, and colonization of, isolated habitat patches (Demers et al. 1995; Haas 1995; Wegner and Merriam 1979). Such plantings are used by a wide array of animals from insects to mammals and provide short-term shelter, foraging sites, and occasionally breeding sites. Benefits are greatest for generalist species and those that occupy scrubby and early successional habitats (Warkentin et al. 1995). Generalist species comprise some of the most notorious mesopredators, however, including nest robbers such as raccoons, skunks, jays, crows, and magpies not to mention exotic species such as starlings. Nevertheless, species that naturally occupy mature woods, forest, or late successional habitats will utilize artificial plantings if the plant-species composition is appropriate (M. S. Foster, unpublished data) and the density of food resources sufficient (Greenberg et al. 1995; Russell et al. 1994).

The second level of connection occurs at the landscape mosaic scale. Corridors at this scale are broader, longer, and designed to connect major landscape features such as woodlots, wetlands, or adjacent peaks within a mountain range. Their function is to allow the daily movement or seasonal dispersal of species that are normally restricted to certain types of habitat—for example, forest-interior bird species. They also provide substantial areas of breeding habitat for edge species.

The regional scale is the largest level of connection suggested by Noss (1991). Wildlife corridors at this scale reconnect large areas of land that would otherwise become isolated. Such connectors also contain significant areas of land that might be nature preserves in their own right. As conceptualized by The Wildlands Project, systems of core habitat and regional corridors are the basis of a plan to connect national parks, wilderness areas, and other wildlands, existing or potential, on a continental scale. The Yellowstone-to-Yukon Project is an example of the need for connections at the regional scale. It seeks to ensure connectivity between northern Canada and the Greater Yellowstone Ecosystem, facilitating the dispersal of grizzlies and other large carnivores in the region. Connections are being designed that will link the Yellowstone region to Idaho and Glacier National Park, which in turn have connections to Banff and Jasper and beyond.

If we take the perspective of the animals themselves, however, we come up with a slightly different typology of movement in space—one that emphasizes species-specific activities. Consider foraging. A shrew forages on a smaller scale than does a wolf. For a shrew, all movements are local (or at the hedgerow scale) based on the movement categories described earlier. For a wolf, however, foraging occurs at the landscape mosaic or regional scale. For both species, daily activities such as foraging, predator escape, and mate finding take place within an individual's home range. The point is to view the world's natural connections through the eyes, ears, and noses of a range of other species. The human perspective will always be a distorted subset of those employed by other species.

## Dispersal

Dispersal occurs when individuals move out of the area occupied by a family group or local population. In birds and mammals, *natal dispersal* occurs when young animals disperse from their birth site to the site where they first breed; *breeding dispersal* is the movement between breeding sites in successive years (Greenwood 1980). Dispersal is an innate behavior, but rates of dispersal are sensitive to local resources (food, shelter, space) and the number of individuals utilizing these resources. Resource deficiencies occur seasonally. In many species they usually precede the reproductive period, so that dispersal of juveniles attempting to establish their own territories is usually accompanied by a significant increase in mortality. Dispersal of adults is likely to be enhanced considerably when environmental perturbations such as drought, floods, or fire cause declines in available resources. Dispersing animals may move short distances to occupy recently vacated adjacent territories in the same population area, or they may move long distances and enter other populations or settle in unoccupied and marginal areas. Many animal species exhibit "minimum dispersal distances"—especially during juvenile dispersal. Most individuals will move for a considerable distance, or time, before settling or establishing their territories. In contrast, the social systems of many species have evolved due to lack of opportunities for dispersal and establishment of an individual territory. The creation of unnatural dispersal corridors—or the artificial translocation of key individuals—may disrupt the social system in species such as lions, elephants, and many primates.

For some species, radiotelemetry data can be used to estimate the frequency distributions of dispersal distances achieved by individuals. These distributions are always skewed: most animals move short distances, but a few move over longer distances. (The observed distribution is in turn biased by the difficulties associated with locating animals that move large distances.) Although this information is essential for designing wildlife corridors, it is dangerous to rely on mean or modal dispersal distances. Instead one should consider the shape of the dispersal distribution.

Caution is needed when using data on animal movements and generalizing about movement over different habitat types. Data on dispersal distances seldom quantify the quality of habitat encountered during movement, making it difficult to guess how reliable these data are for estimates of connectivity. Nonetheless, direct observations of movement can provide estimates of connectivity, particularly when they are available for the routes of interest—for example, Beier (1996) observed only two movement events of pumas across a narrow movement corridor in southern California. When it comes to judging the potential utility of these corridors, these data are far more convincing than any complex ecological model.

Long-distance dispersal provides demographic and genetic exchange between populations while maintaining the potential for the reestablishment of populations in areas from which the species has been extirpated as a result of human or natural disturbance. Short-distance dispersal should be accommodated within core areas of reserves but may be disrupted by the presence of roads, dams, or other built structures. Long-distance dispersal, by contrast, may involve movement of individuals among populations—either within a single core area or among those occupying widely separated core areas in a fragmented landscape.

Even activities that take place wholly within the core area of a reserve may require corridors, assuming some development exists in the core area (as in Banff and many other national parks). In such circumstances, simple structures may allow animals to cross roads safely or expedite movement around dams. Short-distance connectors include such structures as highway underpasses or overpasses, frog tunnels, and fish ladders (Langton 1989).

## Large-Scale Movements

Some animals are nomadic for most of their lives—for example, wolverines and crossbills move constantly in a seemingly erratic fashion through a habitat or geographic area in response to irregular patterns of food availability (Adkisson 1996; Benkman 1992). Migrants, by contrast, make regular, periodic movements between two distinct areas—for example, breeding and wintering areas or seasonal feeding areas (Roca 1994). Movements may be elevational, latitudinal, or both (Powell and Bjork 1995). Nomads and migrants tend to cover extensive geographic areas that are generally too large to be incorporated into a single core area— for example, caribou in Alaska, wildebeest in the Serengeti, and saiga in Outer Mongolia. Some species, such as elk, migrate from uplands to lowlands in the fall and go back to the mountains in the spring. Others, such as golden plovers and arctic terns, cover thousands of kilometers during their transoceanic migrations.

Elevational migrations of animals at different seasons are significant—particularly in the tropics (Crome 1975). Nevertheless, corridors that link habitats at different elevations in the tropics have been little studied. A notable exception is the study by Janzen (1983), who described the elevational migrations of adult lepidoptera in northwestern Costa Rica. These lepidoptera, such as *Xylophanes turbata* (Sphingidae), become abundant in the lowland forest during the early rainy season. Females lay their eggs at this time. The resulting large populations of larvae cause extensive defoliation of their host plants (*Psychotria microdon*, Rubiaceae, in the case of *X. turbata*). The caterpillars produce a new generation of adults two weeks later. The new adults are evident for a few days and then disappear for some ten months; evidently they migrate to cooler, high-elevation, evergreen forests where metabolic costs would be lower. Apparently the adults reproduce at high elevations as well. Several other insects exhibit similar elevational movements. Many of these play key ecological roles—including leaf eating as larvae and pollination as adults. Some are also important prey for other animals. For these elevational migrants, and for the species that interact with them, vertical connectivity is absolutely critical.

A contrasting case—in terms of spatial extent—is that of the North American subspecies of little long-nosed bat (*Leptonycteris curasoae yerbabuenae*). It occurs from the southwestern United States to El Salvador. Northern populations of this bat are migratory, performing seasonal long-distance movements from southern Arizona and New Mexico to tropical and subtropical areas of the Mexican states of Sonora, Sinaloa, and Durango (Cockrum 1991; Wilkinson and Fleming 1996). In central Mexico, some populations of long-nosed bats perform altitudinal movements from the lowland tropical dry forests and arid shrublands to mid-elevation temperate forests (Herrera-Montalvo 1997). Other populations remain in the same sites year-around (Ceballos et al. 1997).

Long-nosed bats are specialized nectar feeders, although they complement their diets with fruit. Close mutualistic relationships have been documented between long-nosed bats and plants such as century plants and related forms (*Agave* spp., *Manfreda brachystachya*), silk trees (*Ceiba, Pseudobombax*), and columnar cacti (Howell 1974; Eguiarte et al. 1987; Fleming et al. 1993; Valiente-Banuet et al. 1996). Because of their specialized diet, long-nosed bats require a continual supply of blooming plants along their migratory routes. Therefore, any widespread loss of habitat along this "nectar trail" (Gentry 1982) could disrupt the entire migration of bats (Fleming et al. 1993) and spell disaster as well for the species they pollinate. A complicating factor is that migration is coupled with reproduction in long-nosed bats, for pregnant females travel each year to give birth in the caves of northern Sonora and Arizona (Cockrum 1991). Because of its intricate nature, therefore, the migration of long-nosed bats can be considered an "endangered phenomenon" in need of protection. (See Brower and Malcolm 1993; Nabhan and Fleming 1993; Arita and Prado in press.)

Migratory movements of long-nosed bats demonstrate the importance of connectivity in designing strategies for the protection of biodiversity—not only for the bat itself but also for the maintenance of interactions and ecological processes in several ecosystems that may seem disconnected to the casual observer. We see that habitats that are seemingly unrelated (such as the Arizonan desert and the tropical forests of northern Mexico) are indeed connected by animals, such as the long-nosed bat, that act as "mobile links." Habitat loss in Arizona could have a detrimental effect on the populations of silk trees in distant Sinaloa if populations of long-nosed bats (the main pollinators of the silk trees) are affected. In a similar fashion, the bats connect lowland semiarid areas of central Mexico and mid-elevation temperate habitats.

Several species of birds—including important pollinators and seed dispersers such as resplendent quetzals (*Pharomarchus mocinno*), oilbirds (*Steatornis caripensis*), and bellbirds (*Procnias tricarunculata*)—perform latitudinal or altitudinal migrations similar to those of the long-nosed bats. Similarly some butterflies, particularly the monarch (*Danaus plexippus*), are capable of long-distance movements through several types of habitat. A comprehensive strategy for the conservation of these habitats should take into account the natural connectivity between them. A system of protected areas, to be effective, should preserve this connectivity by assuring the conservation of natural habitat along the migratory routes of keystone species. Later we address the issue of planning for the needs of migratory birds.

# Gene Flow: The Genetic Argument for Corridors

How much connectivity is enough to permit sufficient dispersal to ensure no loss of genetic variability? Or more accurately, if we ignore for the moment the many ecological arguments for dispersal, what is the optimal rate of individual movement between populations from a genetic standpoint? In this special context, the purpose of dispersal is to maintain genetic variability, thereby reducing the probability of inbreeding and reduced fitness of individuals and populations. The reader should be warned, though, that the genetic guidelines discussed here may vastly underestimate the rates of movement needed to maintain optimal densities of migrants for ecological interactions and the maintenance of ecosystem diversity and function.

To understand the current views on genetic connectivity in a human-fragmented landscape, it is necessary to consider the history of concepts evaluating connected and unconnected populations. Since the pioneering work of Sewall Wright in the 1930s (Wright 1931, 1932), population geneticists have analyzed the genetic structure of populations across a landscape, incorporating the dynamics generated by different population sizes and levels of connectivity generated by different rates of gene flow. If we apply these theoretical concepts to conservation, we learn that, relative to populations with high gene flow, sets of isolated populations—if they survive—may retain higher levels of allelic diversity because of local adaptation and random genetic drift (see Slatkin 1985; Varvio et al. 1986). Yet each isolated subpopulation loses heterozygosity because of genetic drift, thereby becoming more vulnerable to extinction (Gilpin and Soulé 1986; Mills and Smouse 1994; Frankham 1995; Newman and Pilson 1997). This interplay between perpetuating genetic divergence among subpopulations while minimizing loss of heterozygosity within subpopulations has led to a surprisingly robust genetic rule of thumb: A migration level of one to ten individuals every generation between subpopulations of a species will minimize the loss of heterozygosity within populations (Mills and Allendorf 1996). This guideline is known as the "one migrant per generation" (OMPG) rule (Wright 1931).

Application of the OMPG rule to conservation problems (Frankel and Soulé 1981) has been criticized (Varvio et al. 1986). Even so, the only direct experimental evaluation of the rule (Spielman and Frankham 1992) found strong increases in reproductive fitness in nineteen replicate inbred lines of *Drosophila melanogaster* that received only a single migrant per generation. Similarly, Mills and Allendorf (1996) found that the rule holds up rather well to real-world complexities and constraints such as overlapping generations, pattern of population subdivision, number of subpopulations, and demographic patterns (see also Hedrick 1995). Thus, despite its shortcomings, the OMPG rule remains the obvious starting point for case-specific evaluations of desirable levels of genetic connectivity. Although the OMPG guideline also applies in principle to fragmented populations of any size, a series of important caveats applies here. In particular we should bear in mind that the OMPG addresses only genetic concerns and ignores such factors as levels of inbreeding or out-breeding (Leberg 1990), demographic/environmental variation (Brown and Kodric-Brown 1977; Lande 1988), disease transmission (Wilson et al. 1994; Hess 1994), and behavioral disruption (Soulé 1983). Each of these factors may cause a different number of migrants per generation to represent an ideal level of connectivity.

It is possible, of course, to have too much connectivity. Corridors may expose wildlife to diseases of domestic animals, and local populations may be overwhelmed by immigrants. But too little connectivity can lead to inbreeding depression within populations and an increased chance of extinction due to environmental or demographic variation. How do we decide what level of connectivity is just right for a situation? Although the answer to this question is often idiosyncratic, depending on the particulars of each situation at a specific point in time, it is worth exploring the possibility of general guidelines. The OMPG rule is the only operational rule currently available, but it is not entirely satisfactory because it is based purely on genetic considerations. Another approach that is equally viable on theoretical grounds is to estimate optimum levels of dispersal for creatures with different life histories (Olivieri et al. 1990; 1995). This approach may imply that higher levels of dispersal are required for species distributed as metapopulations to persist. This conclusion replicates Lande's (1988) implying that by the time a population is small enough for inbreeding depression to be a significant threat, the population is likely to be in desperate straits for purely demographic reasons.

Instead of searching for a universal rule, perhaps we can derive an appropriate level of connectivity from the animals themselves. For example, perhaps we could use radiotelemetry or capture/recapture studies (Ims and Yoccoz 1997) to determine the rate of connectivity or movement across an unfragmented landscape—such as a wilderness area—and make that our standard rate of connectivity in a fragmented landscape. Although there may be compelling reasons for more or less connectivity than this standard, it could provide an operational starting point. Despite its conceptual appeal, there are several logistical problems with this approach. First, for most species in most areas there are no longer unfragmented "control" areas to provide such a baseline. Second, radiotelemetry (and especially trapping) generally underestimates movement rates and distances due to the difficulty in detecting rare but important dispersal events (Koenig et al. 1996). Furthermore, knowing that an animal moves is only part of the picture: the immigrants must reproduce (or facilitate reproduction by others) before they contribute to the demographic or genetic benefits of connectivity.

The conundrum of determining how current levels of connectivity compare to the past might be illuminated by emerging molecular techniques. Genetic tools have the benefit of measuring connectivity from the viewpoint of individuals that matter most: those that breed and add their genes and offspring to the population. The most straightforward way to compare past and current levels of connectivity is to compare genetic variation in samples collected prior to fragmentation to those collected under current fragmented conditions. In a simplified sense, isolated populations lose genetic variation—so that quantifying the loss over time provides insights into the degree to which connectivity has declined. Thanks to recent advances in analysis of DNA using the polymerase chain reaction, we can analyze very old samples from wall mounts or museum specimens (Hagelberg et al. 1991; Mundy et al. 1997) and can also collect contemporary samples noninvasively (without having to capture the animal) from hair, feces, urine, and so on (Morin and Woodruff 1996; Kohn and Wayne 1997; Schwartz et al. 1998). Thus levels of genetic variation in prefragmentation samples can be compared to current samples (Bouzat et al. 1998).

It is even possible to learn about past levels of connectivity just by looking at the current population structure of a variety of species across a fragmented landscape. Isolated populations evolve independently, and the more isolated populations become more genetically differentiated. As the level of differentiation can be measured with genetic analyses, the degree of movement leading to that differentiation can be inferred. (See Avise 1994; Slatkin 1994, 1995; Goudet 1995; Templeton and Georgiadis 1995; Paetkau et al. 1998.) Time must pass for a certain level of connectivity to manifest as a certain level of genetic differentiation. This time to equilibrium depends on connectivity, population size, mutation rates, selection pressures, and other factors, but it could be 100 generations or so (Crow and Aoki 1984; Varvio et al. 1986; Slatkin 1994). Thus samples collected now may provide a mirror into a past when levels of connectivity were different than they are now. Because different ways of analyzing genetic data have different times to equilibrium, we can quantify connectivity at different points in the past. At the most recent part of the continuum, a new class of genetic analysis techniques, called the assignment test, can quantify current rates of dispersal (Waser and Strobeck 1998). Favre et al. (1997) used the assignment test to show the extent to which female shrews disperse more than males.

Genetic-based techniques have already begun to contribute important conservation insights into population structure and connectivity decision making for a wide variety of species ranging from frogs (Hitchings and Beebee 1997) and lizards (Cunningham and Moritz 1998) to wolves (Roy et al. 1994; Forbes and Boyd 1996) and bears (Taberlet et al. 1997; Paetkau et al. 1995, 1998). Are these genetic approaches accessible to those involved in on-the-ground conservation? Both the genetic techniques and analytical tools are being developed rapidly, and measures of genetic variation, population structure, and connectivity are being published every week for a wide array of species. This trend will surely continue. Quantifying movement between populations is an enormous technical challenge (Ims and Yoccoz 1997), but genetic tools are likely to yield revolutionary insights into current and historical movement rates.

In short, genetic considerations have provided an initial rule of thumb for a level of connectivity and empirical studies have begun to identify general factors that should be considered when evaluating the efficacy of a particular kind of connectivity. These factors can be used to decide whether a proposed level of connectivity permits an appropriate level of gene flow or can help us select the most promising corridor or modified matrix or buffer zone.

## Do Corridors Work?

We have already explored the usefulness of corridors. Here we survey empirical studies that address whether increased connectivity can enhance the persistence of populations. Several studies have shown that movement between populations reduces the probability of overall extinction. In these studies, movement reduces demographic and environmental stochasticity and increases the recolonization of extirpated patches (Forney and Gilpin 1989; Blaustein 1981; Wahlberg et al. 1996). Moreover, movement between small, inbred populations has reduced the chance of extinction due to inbreeding depression (Spielman and Frankham 1992; Newman 1996). The question of efficacy is reviewed by Beier and Noss (1998).

Ecosystem diversity and integrity also benefit from the maintenance or restoration of connectivity. Leigh et al.'s (1993) studies on islands in the Panama Canal illustrate that island tree communities become dominated by tree species whose seeds are favorite food items of herbivores on the mainland. As different herbivores have preferences for seeds of different species, and as different herbivores are lost from different islands, the tree community becomes impoverished in different ways and at different rates on islands of different sizes and degree of isolation (see Chapter 3). Kreuss and Tscharntke (1994) found that fragmentation of forest habitats led to a decline in the diversity of insect parasitoids. This reduction in the density and diversity of insect predators led to bigger and more rapid outbreaks of pest insects in the surrounding agricultural landscape. Thus corridors such as hedgerows in agricultural landscapes are likely to provide habitat for pollinating species and for the parasitoids that are the natural control agents of many agricultural pests. Similar phenomena are noted in studies of decomposers. Klein (1989), studying dung beetle communities in Amazonian tropical forest fragments, found that reduced diversity of these communities in smaller and more isolated patches led to reduced rates of dung decomposition and significant changes in nutrient processing.

The question of whether particular corridors function as hoped is much more difficult to establish. Studies have shown that animals use linear patches either as movement corridors or as additional habitat (see Rosenberg et al. 1997), but the extent to which corridors actually increase the rate of successful movement and establishment has not been determined. Furthermore, the question "Do corridors work?" may be sterile because the answers are indubitably yes, no, maybe, or sometimes. Efficacy depends on the characteristics of the corridor or matrix, the species, the individuals moving, time of year, degree of movement required to connect populations, and many other factors.

Nevertheless, promoting the movement of individuals among the remnant fragments may increase population persistence and local species survival (Fahrig and Merriam 1994). Rosenberg et al. (1997) found that when the contrast between a patch or corridor habitat and its surrounding matrix of human-modified landscapes was strong, greater numbers of salamanders (*Ensatina eschscholtzii*) used corridors to move between patches than crossed the matrix. Although animals crossing the matrix moved faster than those using corridors, they died more readily. When the contrast between corridor and matrix was small, however, the faster movement of animals across the matrix compensated for the cost of being in the matrix and corridors had no positive effect on mortality.

Defenders of landscape connectivity in general and wildife corridors in particular sometimes fall back on the naturalistic premise that the preagricultural world was naturally connected and of a piece. Hence the natural situation is connectivity, not fragmentation. In restoring connectivity, therefore, we are merely repairing the ecological insults of agriculture and development. It follows that the burden of proof should be on those who oppose corridors, not those who support them. Such an argument is too facile, however, for there are times when the restoration or imposition of connectivity may be more harmful than fragmentation. We discuss some of these circumstances later in the chapter.

# Planning: Demographic and Ecological Models

The planning of landscape connectivity is still as much art (including eco-logical intuition) as science. Fortunately, though, not all species require connectivity at the regional level, so the actual siting of corridors is gen-erally based on the needs of just a handful—usually the large, nonflying, wide-ranging animal species. Several modeling approaches are currently being used to determine the best routes for target species, but these quantitative approaches are often inadequate for lack of data or preci-sion. In the meantime, however, opportunities for reconnecting wild-lands are slipping away, so a combination of quantitative and qualitative approaches will continue to be used and refined. Among the most valu-able of the available sources of information, given the inherent com-plexity of nature, is the intuition of experienced field biologists.

## Individual Movement

It should be apparent that the first step in the analysis of corridor capability is the selection of target species (Soulé 1991). Corridors, like any other conservation element, require justification. Above all, corridors must be justified in terms of their ecological functions. There-fore, the idea of a generic landscape corridor—connectivity for the sake of connectivity—is more aesthetic than scientific and will generally be dismissed in the hard light of scientific review. Thus the first stage in corridor analysis should always be a study to determine what species and what essential ecological processes require connectivity. The pri-mary candidates for corridors are species that are not viable in the long term in the core reserves or species that are nomadic or migratory. In most cases, then, small, abundant species need not be considered—par-ticularly when one is working at a regional or landscape scale. The most likely candidates, therefore, are relatively rare, wide-ranging species such as large carnivores (Soulé 1991) and perhaps late-successional tree species.

The second stage of corridor analysis is to determine the utility of a particular route for the selected corridor users. A corridor that links mountain ranges across scrub or steppe lowland may work well for mountain lions (*Felis concolor*), for example, but is probably useless for marten (*Martes americana*) or wolverine (*Gulo gulo*). Each situation must be evaluated to determine whether there are species or processes that require such links. Assuming, then, that our goal is to maintain connec-tivity of certain species across a landscape, and assuming that we know everything relevant about the landscape, is it possible to create a model that produces the best outcome for connectivity? In theory, yes. And in well-defined situations the results for species like the grizzly bear may make sense intuitively. In practice, however, it is difficult.

The emergence of island biogeography has demonstrated that both the size of the patches and the distance between them affect levels of movement (MacArthur and Wilson 1967). In turn the degree of connectivity affects colonization and extinction rates (MacArthur and Wilson 1967; Brown and Kodric-Brown 1977). Early application of these principles to conservation problems (Diamond 1972; Terborgh 1974) led to a recognition that evaluations of connectivity must embrace the biophysical nature of the connecting matrix as well as the biology and behavior of individual dispersing species (see Bolger et al. 1991; Taylor et al. 1993; Doak and Mills 1994).

Two general approaches are used to predict movement in a landscape and its consequences: partial differential equation models (PDE) of population dynamics, including spatial flow (Skellam 1951; Okubo 1980; Holmes et al. 1994; Turchin 1998), and simulation models of individual dispersers. Over the past twenty years, PDE models in ecology have been highly elaborated and now incorporate many of the complexities of dispersal behavior and landscape pattern. (See Turchin 1998 for a comprehensive review and synthesis.) Pollen movement by insects, for example, can be successfully modeled to predict genetic flow connectivity across heterogeneous landscapes (Morris 1993). Nevertheless, there are two reasons why PDE models are unlikely to be reliable tools for investigation of large-scale connectivity questions. First, PDEs assume a continuous distribution of population densities and are therefore poor at predicting processes involving only a few individuals. Predicting conservation connectivity, as noted earlier, usually means estimating whether a few members of a rare species will move large distances. Second, PDEs that incorporate complex spatial landscapes are not generally amenable to analytical solutions and therefore lose the major advantage of the PDE approach over simulation modeling.

The second approach—simulation models of movement and connectivity—range from quite simple (Doak 1989) to very complex depictions of individual behavior and landscape characteristics. As computing power has grown, the trend in simulation modeling has been toward more and more complex models of movement and behavior. Above all, the ability to overlay simulated dispersers onto GIS landscape maps has greatly increased the realism of spatial simulations (Liu et al. 1995). Simulations have been used both to assess general landscape characteristics that will promote connectivity (Doak et al. 1992) and to assess the adequacy of certain landscapes for particular species (With 1997).

While it is clear that simulation models are powerful tools for assessing landscape connectivity, their limitations are often overlooked. The more realistic are governed by parameters, but there are seldom enough data. While many studies employ informal sensitivity analysis, few have tested the effects of their model's structure and parameter values on predictions of landscape connectivity or population health. Thus we have little hope of knowing whether conclusions based on these models are optimistic or pessimistic. A recent study by Ruckelshaus et al. (1997) shows that estimates of connectivity can be highly sensitive to the exact parameter values governing individual behavior. Hence the exact results of simulation studies must be viewed with extreme skepticism. Planners should be cautious when building models whose complexity demands data that do not exist. Instead, they might consider more abstract simulations, such as those based on percolation theory (With 1997) or the assessment of landscape connectivity in a general way (Schumaker 1996; 1998).

Some of the most powerful approaches to assessing connectivity are not nearly so complex as those just described—and simplicity is a virtue that is usually overlooked. Lande's (1988) analysis of spotted owl movement between potential habitat patches avoids any explicit consideration of spatial processes and ignores landscape patterning completely. Yet it constitutes the most influential analysis of landscape connectivity undertaken to date. Straightforward models that make the minimum number of sensible assumptions may be the best way to begin analyzing landscape connectivity.

## The Complication of Multiple Habitats

The issue of connectivity is most straightforward when we can distinguish between suitable and unsuitable habitats. But as the debate over the usefulness of habitat corridors suggests, corridors and other nonideal habitats may in fact support at least some resident populations, sometimes acting as demographic sinks (Pulliam 1988). While most of the modeling approaches sketched here incorporate habitats of varying qualities, variable habitat quality complicates the estimation of the costs and advantages of different connectivity solutions. If we are assessing the role of a patch of native grassland as a stopover point for migrating mice, for example, we need to know a great deal about the dispersal behavior of the species but not much about mice demographics. If this restored area may entice dispersers to take up residence, however, we also need to know the demographic rates that will govern the mouse population on this small habitat patch—and how, in sum, the patch will affect dispersal and demography. In short, when alternative schemes for connectivity will influence demographic rates other than the mortality incurred during movement, the problem of assessing different landscape patterns and plans becomes even more complex.

The common modeling approaches may be much better at determining relative degrees of connectivity than setting minimum standards for connectivity. That is, if we use these approaches to choose between different restoration and preservation options—rather than set minimum standards for interpatch distances or corridor widths—they may give reliable guidance for large-scale planning. More work is needed before we can undertake quantitative analysis of landscape connectivity without debilitating caveats. Moreover, there is still a role for the experienced wildlife biologist or naturalist who knows the target animal and what path it is likely to take in a fragmented landscape. For the time being, corridor design is as much art as science.

## The Notion of Metapopulations

Recently, the consequences of connectivity between populations of species or groups of species have increasingly been viewed in the context of "metapopulations." (See Hastings and Harrison 1994; Hanski and Gilpin 1991; Hanski and Simberloff 1997.) After exposure to a battery of field and experimental tests (Harrison et al. 1988; McCullough 1996; Wahlberg et al. 1996; Hanski and Gilpin 1991; 1997), this concept has been subsequently broadened and revised. Originally it referred to systems of isolated population units that periodically go extinct ("blink out") and are reestablished ("blink on") by dispersing individuals from other units. Now the concept is used to denote almost any system of populations whether or not they blink out periodically.

The fact that most species in nature are not as structured as metapopulations in the original sense (references in Hanski and Simberloff 1997) has not prevented popularization of the term, particularly among nonscientists. Harrison (1994:117) summarizes the current status of the metapopulation concept as a basis for our thinking about connectivity: "It seems necessary to adopt a broader and vaguer view of metapopulations as sets of spatially distributed populations, among which dispersal and turnover are possible but do not necessarily occur. Such a definition leaves little hope for strong generalizations about the role or importance of metapopulations. A possible way forward is to ask, in each specific case, 'what is the relative importance of among-population processes, versus within-population ones, in the viability and conservation of this species?'"

## Corridor Controversies

The advent of interest in movement corridors for conservation coincided with the rise in popularity of the dynamic equilibrium theory of island biogeography (MacArthur and Wilson 1967). Within a decade, a series of papers applying island biogeography theory to refuge design led to the widespread acceptance of a series of rules for such endeavors. These rules became the governing paradigm in conservation biology (Hanski and Simberloff 1997). One rule states that a set of refuges connected by corridors will preserve more species than the same set of refuges without the corridors. The underlying reasoning is that immigration into each component refuge will raise its equilibrium number of species and "rescue" populations that are inviable for demographic or genetic reasons. This reasoning does not directly address the issue of whether the entire network will thereby contain more species. Nor does it address whether the corridors were intended simply for movement or as breeding habitat as well.

The initial popularity of movement corridors (Hudson 1991; Saunders and Hobbs 1991; see references in Simberloff et al. 1992; Hussey et al. 1989) seems to have resulted from a combination of attributes: insufficient habitat in core reserves in many regions of the world; the concept's inherent attractiveness and simplicity; and the idea that purchasing movement corridors is cheaper than buying huge amounts of habitat. But even if corridors work, it does not always follow that they are the best solution. Scale issues are relevant here. Much of the debate has focused on connectivity at the intermediate landscape mosaic scale. While this debate has been instructive, it has also created a misconception about the requirements for viability of species and the integrity of ecosystems within a landscape. Obviously many key ecosystem functions—such as pollination, nitrogen fixation by mycorrhizae, nutrient decomposition, and recycling—take place on a small, local scale. Other ecosystem functions, by contrast, require large areas of continuous habitat to function efficiently. Among the most important are the maintenance of vigorous carnivore populations. Fragmentation also reduces the capacity of ecosystems to store nutrients, sequester carbon, and provide pest protection. For example: water storage and recycling require large patches of both forest and wetlands; forest cover minimizes soil erosion when rain is falling; wetlands and lakes provide long-term storage and slow release of water between wet seasons or during long-term droughts.

Migratory ungulates, too, also require connectivity. The large-scale antelope migrations—such as wildebeest in the Serengeti, elk and caribou in North America, and saiga in southern Mongolia—serve two important functions. First, migrations allow the herds to escape the buildup of biting insects and other parasites that would occur if the species were sedentary. Second, although these seasonal movements may have evolved to avoid pests, they also create the opportunity for the vegetation to recover from grazing in different parts of the species' range—leading to a more equitable use of forage throughout their range. In areas where the habitat is fragmented and the traditional migratory routes are interrupted, both overgrazing and increased levels of attack by biting flies and parasitic helminths result. This is detrimental to both the herbivore and its habitat. In summary, scale is an important factor when deciding whether to invest in connectivity. Where large, wide-ranging species are at risk, some kind of connectivity, existing or restored, is essential.

But what are the costs of connectivity? It is critical to weigh the potential benefits of creating connections against likely costs in order to assess the practicality of such conservation plans. In most cases, benefits will be realized only if populations inhabiting fragments are indeed too small to persist for centuries in their present state, if the proposed corridor will significantly improve this situation, and if the corridor does not become an avenue of flux between sources and sinks—thereby sending the entire interconnected unit into decline. Unless these conditions are met, corridors often cost more than they are worth. But even if a proposed corridor is necessary and sufficient to unite populations and habitats that are incapable of persisting in isolation, at least six classes of potential costs must be considered. The most important of these, usually, are trade-off costs.

First, trade-off costs exist when a fixed allotment is available for the project so that any investment of funds, time, or human resources in establishing corridors causes an equivalent reduction in what is left for other things. The most common example is the trade-off between the cost of corridors and the cost of enlarging existing habitat or population units.

Second, there are costs associated with maintenance of connectivity. Such costs include the construction and repair of fences or other barriers and the protection of wildlife while using the corridor. Protection is especially important when animals using the connector are vulnerable to human activity. These costs must be judged in comparison with the costs necessary to maintain the core reserves. Land used for corridors or stepping stones may be unsuitable habitat for the species of concern due to its past or present use. When this is true, restoration costs must be added—and such costs often become substantial.

Third, corridors often contain much edge habitat and hence are vulnerable to deleterious edge effects. Studies of the California red-backed vole (*Clethrionomys californicus:* Mills et al. 1995) and several bird species (Paton 1994), for example, show strong negative edge effects—leading to predictions of reduced survival in linear, high-edge environments (Simberloff et al. 1992). Beier (1993, 1996), however, argues that even low levels of connectivity in low-quality corridors may be better than none at all. Another investigation of corridor types (Andreassen et al. 1996) has shown that corridors which are too narrow may be avoided by root voles (*Microtus oeconomus*) whereas those which are too wide may make travel inefficient and slow, exposing the animal to predation (Soulé and Gilpin 1991; Rosenberg et al. 1997).

Fourth, corridors may have political costs as well. These costs accrue because increased connectivity nearly always requires land-use patterns to be altered, thus forcing people who are using the land, even public land, to give it up or change their behavior. Of course, this conflict can also occur over the enlargement of existing habitats or population fragments. Corridors often necessitate acceding land of more varied uses, however, thus increasing the likelihood that the costs of giving these up will aggravate someone. A common result is that some element of society becomes angered and hence less sympathetic to the conservation project and indeed conservation in general.

Fifth, the last ten years have produced increasing evidence that global climate change will present a major threat to the preservation of biodiversity. Certainly the habitat connectivity that permitted long-range movement of species in response to previous global climate changes (Davis 1981; Coope 1995) has been destroyed by anthropogenic habitat fragmentation. There is widespread uncertainty that even massive corridor systems would greatly ameliorate the impact of global warming, but many of those voicing doubts (Hobbs and Hopkins 1991) advocate establishing these corridors anyway. To be useful in this context, corridors would have to be wide enough and contain habitat of sufficient quality to ensure that feeding and breeding as well as movement would be possible.

And sixth, corridors may allow infectious diseases to spread between one section of a population and another (Hess 1994). Similarly, connections across the landscape may allow fires or insect pests to spread between patches of forest that would otherwise be isolated. Yet pathogens, pests, and even fires are among the natural factors that determine the age, size, or immunological structure of a species. In many cases preventing a pathogen or fire from threatening a species or ecosystem may produce the potential for a much more serious disaster (Dobson and May 1986a; McCallum and Dobson 1995). In a population that experiences regular outbreaks of a pathogen such as malaria or schistosomiasis, for example, most individuals exhibit some level of immunological resistance. This resistance creates "herd immunity"—that is, the presence of a significant proportion of immune (and partially immune) hosts significantly reduces the potential for a large-scale epidemic that might be devastating to the host population (Anderson and May 1991; Woolhouse et al. 1991). Where host populations are isolated and disease exposure is less frequent, however, levels of herd immunity are considerably reduced as the population consists mainly of susceptible individuals. In this situation the potential for epidemic outbreaks is considerably enhanced.

In human populations, devastating epidemics often occur when usually benign childhood diseases, such as measles, are introduced into island populations where levels of herd immunity are low and individuals acquire infection for the first time at an older age than usual. Thus connections within the landscape have another function that is initially counterintuitive: they permit the normal flow of common infectious diseases and pathogens which ensures that levels of herd immunity are maintained. Where corridors are adjacent to agricultural land, however, this creates the potential for migrating and dispersing hosts to acquire novel pathogens from agricultural hosts or species with which they rarely have close contact (Dobson and May 1986b). Under these circumstances, simple corridors may aid the spread of novel pathogens. The simplest way to prevent this is to preserve large natural connections in the landscape or ensure that they are restored in a way that minimizes contact between dispersing wild species and sedentary domestic species—a goal that may be very difficult to attain in practice.

## Alternatives to Corridors

Not every species requires designated corridors or physical continuity of wildlands throughout a region, let alone a continent. Nor does every ecological process require connectivity on the ground or in the water. In this section we discuss a variety of situations that permit or require as much planning and management as do actual physical corridors.

## Artificial Transport

What are the alternatives to corridors? In some instances, capture and artificial translocation of animals (for example, by truck) may be cheaper and more efficient. Long-distance dispersers and especially migrants that use both aerial and terrestrial habitats tend to range over geographic areas far beyond the spatial extent of any reserve core and too widely separated to be connected by corridors.

There is an extensive literature on moving plants from one area to another. Although it is generally preferable to achieve connectivity for many species at once, in some cases the conveyance of a few individuals of a species may be cost-effective and achieve sufficient connectivity for genetic purposes.

## Other Forms of Landscape Management

Another alternative is managing the matrix so that animals can easily pass through it, occasionally using it for feeding, breeding, and other purposes (Simberloff et al. 1992). An example of this approach, termed the "new forestry" by Franklin (1989), essentially suggests that although reserves will never be sufficient by themselves, the entire forest landscape of the Northwest, including logged areas, can be made far less inimical to native species.

The idea of "habitat variegation" (McIntyre and Barrett 1992) is a similar plan. In this case the tablelands of northern New South Wales can be managed both for extractive uses and for conservation. Exploitation of large tracts of longleaf pine forest in the southeastern United States may similarly be compatible with conservation if appropriate management procedures are applied (Simberloff 1993). This solution may be practical in areas where land-use guidelines are stable, predictable, and enforceable. Such places, however, are becoming the exception rather than the rule (Soulé and Sanjayan 1998). In most of the tropics, the assumption that there will be habitat outside of strictly protected reserves is dubious at best.

## Stepping Stones

The design of reserve networks to accommodate migratory animals, including many species of birds, bats, and insects, presents special challenges. More than half the species of birds breeding in the United States and Canada migrate to Mexico, Central and South America, or the Caribbean Islands for the winter, for example, but make use of terrestrial stopover sites along the way. Not only do these birds routinely cover hundreds or even thousands of kilometers, but the migration routes of many species are geographically distinct in the spring and fall and may differ as well according to sex and age. It would be impossible, therefore, to protect continuous habitat corridors over the length of the migration routes that connect breeding and wintering areas.

Most flying migrants can be accommodated using isolated habitat patches, or stepping-stone refugia, during migration. Some species, however, such as shorebirds, have stringent site-specific requirements (Myers et al. 1987). The Western Hemisphere Shorebird Reserve Network Project, of Wetlands for the Americas, has focused on just such an effort for the purpose of protecting breeding, staging, and wintering sites of shorebirds. The integrity of the stepping-stone remnants, then, becomes a matter of concern. At the very least they must provide for species that need to rest and replenish energy while moving from one long-term area of residency to another (Moore and Simons 1992; Winkler et al. 1992). We must consider also that the sizes, shapes, and degree of dispersion of a network of stepping stones that would achieve substantial connectivity will differ from species to species and from taxonomic group to taxonomic group.

Stopover reserves, therefore, must be managed—particularly if they are small and subject to external influences. Ideally they must contain sufficient habitat for a substantial number of nonmigratory species in the regional pool. Management challenges are compounded when migrants must cross political boundaries, whether domestic or international, underscoring the need for regional and multinational partnerships to develop and preserve the networks of widely separated breeding, staging, and wintering areas. The neotropical migrant lazuli bunting (*Passerina amoena*) exemplifies the challenges of managing for the seasonal requirements of a wide-ranging migrant. This species breeds almost exclusively in the United States and winters south to southern Mexico. Thus any effort to maintain viable populations of lazuli bunting would require an international effort involving the United States and Mexico, with special attention given to relatively small areas in Baja California del Sur, southern Arizona, New Mexico, and extreme northern Sonora, areas where it is known to molt and is therefore especially vulnerable (Greene et al. 1996).

To protect neotropical migrants like the lazuli bunting, international organizations such as Partners in Flight have been created as joint partnerships involving private, state, province, and national government agencies and organizations. Partners in Flight is a multinational coalition involving more than a dozen countries and hundreds of state agencies, nongovernmental organizations, and private industry in Canada, the United States, Mexico, and Central and South America. It is dedicated to the maintenance of neotropical migrants through protection of habitat and mitigation of threats to migrant birds on wintering and breeding areas and during migration. Another example of such approaches to conservation is the North American Waterfowl Management Plan created in 1986 as a joint venture between the United States, Canada, and Mexico. Recognizing that the habitat needs of waterfowl transcend federal political boundaries and that the landownership (federal, state, and private) of waterfowl habitat is complex, the participating governments have created partnerships with state agencies and private industry to protect and restore some 1.1 million acres of wetlands habitat.

It is clear, then, that we must think about cultural and political differences as well as biological issues when designing continental and intercontinental conservation projects (Foster 1992). If we are to maintain healthy ecosystems that cross political boundaries, multijurisdictional partnerships are needed. The political scale (neighborhood, hamlet, county, province, nation, continent) cannot be ignored.

## The Sense of Connection

The Wildlands Project has set an ambitious agenda for conserving biodiversity in North and Central America. And connecting and expanding the current set of nature reserves and wilderness areas is central to this goal. Critics of The Wildlands Project tend to regard its scale as utopian. Yet every strand of the scientific argument suggests that if we wish to maintain a viable landscape that allows 100 percent of the native species to persist, then the primary goals of reconnecting the landscape make excellent sense ecologically.

The strongest argument is the simplest. Most of the species that evolved in North America, and on all other continents, did so in a landscape that was heterogeneous but extensively connected. Most of the massive radiation of diversity and abundance occurred before humans arrived and proliferated. Indeed, it is the recent and rapid expansion of the human population that has led to such a decline in the distribution and abundance of other species. In the United States this process has occurred in the last four hundred years, more widely in the Americas over the last ten thousand years, and throughout the world in the last fifty to one hundred thousand years. Although overexploitation is responsible for the decline of some species, in most cases the decline is due to the loss and fragmentation of their natural habitats combined with the introduction of alien weed, pest, and pathogen species. It is these indirect effects of human development that have the biggest impact on biodiversity. The crucial way—perhaps the only way—to minimize future impact is to reduce current rates of habitat loss and explore ways of expanding current systems of reserves by creating large-scale connections across the landscape.

There are many ways of achieving connectivity in a system the size of that envisioned in The Wildlands Project. These approaches are not mutually exclusive. Not only is the biotic community different in different regions, but the traditional management of land may dictate different procedures. In many parts of the East, for example, land would have to be purchased and restored to provide corridors for connectivity—and, due to economic and technological limitations, only narrow corridors might be achieved. In the West, however, land prices are lower so larger areas might be purchased to connect the current reserve system together.

Different species, moreover, will achieve connectivity by different means. Even within one taxonomic group, the heterogeneity of behavior and physiology of different species will cause them to perceive different degrees of fragmentation in the same landscape. The Wildlands Project has focused especially on large animals. But plants, small animals, and certain processes (such as fire in a natural fire disclimax) also require landscape connectivity. There is no reason to believe that one approach to providing connectivity will be optimal for all species and processes— unless the scale of the landscape connection is sufficiently large that it functions as an independent ecosystem. Thus conservationists must employ a mix of methods and apply different mixes in different regions. Above all, conservation biologists should focus on conserving fully functioning ecosystems—not simply maintaining viable populations of charismatic species that may or may not be distributed as a metapopulation. And the most viable and most economical way to do this is to connect the landscape in a way that reflects its natural level of connectivity.

# References

Adkisson, C. S. 1996. Red crossbill (*Loxia curvirostra*). In A. Poole, P. Stettenheim, and F. Gill (eds.), *The birds of North America*. Philadelphia: The Academy of Natural Sciences.

Anderson, R. M., and R. M. May. 1984. Spatial, temporal, and genetic heterogeneity in host populations and the design of immunization programmes. *IMA Journal of Mathematics Applied in Medicine and Biology* 1:233–266.

Anderson, R. M., and R. M. May. 1991. *Infectious diseases of humans: Dynamics and control*. New York: Oxford University Press.

Andow, D. A., P. M. Kareiva, S. A. Levin, and A. Okubo. 1990. Spread of invading organisms. *Landscape Ecology* 4:177–188.

Andreassen, H. P., S. Halle, and R. A. Ims. 1996. Optimal width of movement corridors for root voles: Not too narrow and not too wide. *Journal of Applied Ecology* 33:63–70.

Andrewartha, H. G., and L. C. Birch. 1954. *The distribution and abundance of animals*. Chicago: University of Chicago Press.

Anstett, M. C., M. Hossaert-McKey, and D. McKey. 1997. Modeling the persistence of small populations of strongly interdependent species: Figs and fig wasps. *Conservation Biology* 11:204–213.

Arita, H. T., and K. S. D. Prado. In press. Conservation biology of nectar-feeding bats in Mexico. *Journal of Mammalogy*.

Avise, J. C. 1994. *Molecular markers, natural history and evolution*. New York: Chapman & Hall.

Beier, P.. 1993. Determining minimum habitat areas and habitat corridors for cougars. *Conservation Biology* 7:94–108.

———. 1996. Metapopulation models, tenacious tracking, and cougar conservation. In D. R. McCullough (ed.), *Metapopulations and Wildlife Conservation*. Washington, D.C.: Island Press.

Beier, P., and R. F. Noss. 1998. Do habitat corridors provide connectivity? *Conservation Biology* 12:1241–1252.

Benkman, C. W. 1992. Whitewinged crossbill (*Loxia leucoptera*). In A. Poole, P. Stettenheim, and F. Gill (eds.), *The birds of North America*. Philadelphia: The Academy of Natural Sciences.

Bennett, A. F. 1990. *Habitat corridors: Their role in wildlife management and conservation*. Melbourne, Australia: Department of Conservation and Environment.

Blaustein, A. 1981. Population fluctuations and extinctions of small rodents in coastal Southern California. *Oecologia* 48:71–78.

Bolger, D. T., A. Alberts, and M. E. Soulé. 1991. Occurrence patterns of bird species in habitat fragments: Sampling, extinction, and nested species subsets. *American Naturalist* 137:155–166.

Bouzat, J. L., H. A. Lewin, and K. N. Paige. 1998. The ghost of genetic diversity past: Historical DNA analysis of the greater prairie chicken. *American Naturalist* 152:1–6.

Brower, L. P., and S. B. Malcolm. 1993. Animal migration: Endangered phenomena. *American Zoologist* 31:265–276.

Brown, J. H., and A. Kodric-Brown. 1977. Turnover rates in insular biogeography: Effect of immigration and extinction. *Ecology* 58:445–449.

Ceballos, G., T. H. Fleming, C. Chavez, and J. Nassar. 1997. Annual population cycle of *Leptonycteris curasoae* (Chiroptera, Phyllostomidae) at a roost near Chamela, Jalisco. *Journal of Mammalogy* 78:1220–1230.

Chadwick, D. H. 1990. The biodiversity challenge. *Defenders Magazine* 65(May / June):19–30.

Cockrum, E. L. 1991. Seasonal distribution of northwestern populations of the long-nosed bat (*Leptonycteris sanborni*), family Phyllostomidae. *Zoologia* 62:181–202.

Conroy, M. J., Y. Cohen, F. C. James, Y. G. Matsinos, and B. A. Maurer. 1995. Parameter estimation, reliability, and model improvement for spatially explicit models of animal populations. *Ecological Applications* 5:17–19.

Coope, G. R. 1995. Insect fauna in ice age environments: Why so little extinction? In J. H. Lawton and R. M. May (eds.), *Extinction rates.* Oxford: Oxford University Press.

Crome, F. H. 1975. The ecology of fruit pigeons in tropical North Queensland. *Australian Wildlife Research* 2:155–185.

Crow, J. F., and K. Aoki. 1984. Group selection for a polygenic behavioral trait: Estimating the degree of populations subdivision. *Proceedings of the National Academy of Sciences* 81:6073–6077.

Cunningham, M., and C. Moritz. 1998. Genetic effects of forest fragmentation on a rainforest restricted lizard (Scincidae: *Gnypetoscincus queenslandiae*). *Biological Conservation* 83:19–30.

Date, E. M., H. A. Ford, and H. F. Recher. 1991. Frugivorous pigeons, stepping stones, and weeds in northern New South Wales. In D. A. Saunders and R. J. Hobbs (eds.), *Nature conservation 2: The role of corridors.* Chipping Norton, New South Wales: Surrey Beatty & Sons.

Davis, M. B. 1981. Quaternary history and the stability of forest communities. In D. C. West, H. H. Shugart, and D. B. Botkin (eds.), *Forest succession: Concepts and Applications.* New York: Springer-Verlag.

Demers, M. N., J. W. Simpson, R. E. J. Boerner, A. Silva, L. Berns, and F. Artigas. 1995. Fencerows, edges, and implications of changing connectivity illustrated by two contiguous Ohio landscapes. *Conservation Biology* 9:1159–1168.

Diamond, J. M. 1972. Biogeographic kinetics: Estimation of relaxation times for avifaunas of Southwest Pacific Islands. *Proceedings of the National Academy of Sciences* 69:3199–3203.

Doak, D. F. 1989. Spotted owls and old growth logging. *Conservation Biology* 3:389–396.

Doak, D. F., P. Marino, and P. M. Kareiva. 1992. Spatial scale mediates the influence of habitat fragmentation on dispersal success: Implications for conservation. *Theoretical Population Biology* 41:315–336.

Doak, D., and L. S. Mills. 1994. A useful role for theory in conservation. *Ecology* 75:615–626.

Dobson, A. P., and R. M. May. 1986a. Patterns of invasions by pathogens and parasites. In H. A. Mooney and J. A. Drake (eds.), *Ecology of biological invasions of North America and Hawaii.* New York: Springer-Verlag.

Dobson, A. P., and R. M. May. 1986b. Disease and conservation. In M. E. Soulé (ed.), *Conservation biology: Science of diversity.* Sunderland, Mass.: Sinauer.

Downes, S. J., K. A. Handasyde, and M. A. Elgar. 1997. The use of corridors by mammals in fragmented Australian eucalypt forests. *Conservation Biology* 11:718–726.

Eguiarte, L. E., C. M. D. Rio, and H. Arita. 1987. El nectar y el polen como recursos: El papel ecologico de los visitantes a las flores de *Pseudobombax ellipticum* (H. B. K.) Dugand. *Biotropica* 19:74–82.

Fahrig, L., and G. Merriam. 1994. Conservation of fragmented populations. *Conservation Biology* 8:50–59.

Favre, L., F. Balloux, J. Goudet, and N. Perrin. 1997. Female biased dispersal in the monogamous mammal *Crocidura russula:* Evidence from field data and microsatellite patterns. *Proceedings of the Royal Society of London* 264:127–132.

Fleming, T. H., R. A. Nunez, and L. S. L. D. Silveira. 1993. Seasonal changes in the diet of migrant and non-migrant nectarivorous bats as revealed by carbon stable isotope analysis. *Oecologia* 94:72–75.

Florida Greenways Commission. 1994. *Creating a statewide greenways system.* Tallahassee: Florida Greenways Commission.

Forbes, S. H., and D. K. Boyd. 1996. Genetic variation of naturally colonizing wolves in the central Rocky Mountains. *Conservation Biology* 10:1082–1090.

Forman, R. T. T., and M. Gordon. 1986. *Landscape ecology.* New York: Wiley.

Forney, K. A., and M. E. Gilpin. 1989. Spatial structure and population extinction: A study with *Drosophila* flies. *Conservation Biology* 3:45–51.

Foster, M. S. 1992. The international component of managing biological diversity. *Transactions of the North American Wildlife and Natural Resources Conference.* 57:321–329.

Frankel, O. H., and M. E. Soulé. 1981. *Conservation and evolution.* Cambridge, Mass.: Cambridge University Press.

Frankham, R. 1995. Conservation genetics. *Annual Review of Genetics* 29:305–327.

Franklin, J. 1989. Toward a new forestry. *American Forests* (November/December):1–8.

Gentry, H. S. 1982. *Agaves of continental North America.* Tucson: University of Arizona Press.

Gilpin, M. E., and M. E. Soulé. 1986. Minimum viable populations: Processes of species extinction. In M. E. Soulé (ed.), *Conservation biology: The science of scarcity and diversity.* Sunderland, Mass.: Sinauer.

Greenberg, R., M. S. Foster, and L. Marquez-Valdelamar. 1995. The role of the white-eyed vireo in the dispersal of Bursera fruit on the Yucatan Peninsula. *Journal of Tropical Ecology* 11:619–639.

Greene, E., V. R. Muehter, and W. Davison. 1996. Lazuli bunting (*Passerina amoena*). In A. Poole and F. Gill (eds.), *The birds of North America.* Philadelphia: The Academy of Natural Sciences.

Greenwood, P. J. 1980. Mating systems, philopatry and dispersal in birds and mammals. *Animal Behavior* 28:1140–1162.

Goudet, J. 1995. FSTAT (Version 1.2): A computer program to calculate F-statistics. *The Journal of Heredity* 86:485–486.

Guevara, S., J. Meave, P. Moreno-Casola, and J. Laborde. 1992. Floristic composition and structure of vegetation under isolated trees in neotropical pastures. *Journal of Vegetation Science* 3:655–664.

Gustafsson, L., and L. Hansson. 1997. Corridors as a conservation tool. *Ecological Bulletins* 46:182–190.

Haas, C. A. 1995. Dispersal and use of corridors by birds in wooded patches on an agricultural landscape. *Conservation Biology* 9:845–854.

Haila, Y., and O. Jarvinen. 1982. The role of theoretical concepts in understanding the ecological theatre: A case study on island biogeography. In E. Saarinen (ed.), *Conceptual issues in ecology.* Dordrecht, The Netherlands: Reidel.

Hagelberg, E., I. C. Gray, and A. J. Jeffreys. 1991. Identification of the skeletal remains of a murder victim by DNA analysis. *Nature* 352:427–429.

Hanski, I., and M. Gilpin. 1991. Metapopulation dynamics: A brief history and conceptual domain. *Biological Journal of the Linnean Society* 42:17–38.

———. (eds.). 1997. *Metapopulation biology: Ecology, genetics, and evolution.* San Diego, Cal.: Academic Press.

Hanski, I. A., and D. Simberloff. 1997. The metapopulation approach, its history, conceptual domain, and application to conservation. In I. A. Hanski and M. E. Gilpin (eds.), *Metapopulation biology: Ecology, genetics, and evolution.* San Diego, Cal.: Academic Press.

Harestad, A. S., and F. L. Bunnell. 1979. Home range and body weight: A re-evaluation. *Ecology* 60:389–402.

Harrison, R. L. 1992. Toward a theory of inter-refuge corridor design. *Conservation Biology* 6:293–295.

Harrison, S., D. D. Murphy, and P. R. Ehrlich. 1988. Distribution of the bay checkerspot butterfly, *Euphydras editha baynensis:* Evidence for a metapopulation model. *American Naturalist* 132:360–382.

Harrison, S. 1994. Metapopulations and conservation. In P. J. Edwards, R. M. May, and N. R. Webb (eds.), *Large-scale ecology and conservation ecology* Oxford: Blackwell Scientific Publications.

Hastings, A., and S. Harrison. 1994. Metapopulatoin dynamics and genetics. *Annual Review Ecology and Systematics* 25:167–188.

Hedrick, P. W. 1995. Gene flow and genetic restoration: The Florida panther as a case study. *Conservation Biology* 9:996–1007.

Hedrick, A., and S. Harrison. 1995. Gene flow and genetic restoration: The Florida panther as a case study. *Conservation Biology* 9:996–1007.

Herrera-Montalvo, L. G. 1997. Evidence of altitudinal movements of *Leptonycteris curasoae* (Chiroptera: Phyllostomidae) in Central Mexico. *Revista Mexicana de Mastozoologia* 2:116–118.

Hess, G. R. 1994. Conservation corridors and contagious disease: A cautionary note. *Conservation Biology* 8:256–262.

Heywood, V. H. (ed.). 1995. *Global biodiversity assessment.* Cambridge: Cambridge University Press.

Hitchings, S. P., and T. J. C. Beebee. 1997. Genetic substructuring as a result of barriers to gene flow in urban *Rana temporaria* (common frog) populations: Implications for biodiversity conservation. *Heredity* 79:117–127.

Hobbs, R. J. 1992. Role of corridors in conservation: Solution or bandwagon? *Trends in Recent Ecology and Evolution* 7:389–392.

Hobbs, R. J., and A. J. M. Hopkins. 1991. The role of conservation corridors in a changing climate. In D. A. Saunders and R. J. Hobbs (eds.), *Nature conservation 2: The role of corridors*. Chipping Norton, New South Wales: Surrey Beatty & Sons.

Holmes, E. E., M. A. Lewis, J. E. Banks, and R. R. Veit. 1994. Partial differential equations in ecology—spatial interactions and population dynamics. *Ecology* 75:17–29.

Holt, R. D., S. W. Pacala, T. W. Smith, and J. Liu. 1995. Linking contemporary vegetation models with spatially explicit animal population models. *Ecological Applications* 5:20–27.

Howell, D. J. 1974. Bats and pollen: Physiological aspects of the syndrome of chiropterophyly. *Comparative Biochemistry and Physiology* 48:263–276.

Hudson, W. E. 1991. *Landscape linkages and biodiversity*. Washington, D.C.: Island Press.

Hudson, R. R., M. Slatkin, and W. P. Maddison. 1992. Estimation of levels of gene flow from DNA sequence data. *Genetics* 132:583–589.

Hunter, M. L., J. G. L. Jacobson Jr., and T. Webb III. 1988. Paleoecology and the coarsefilter approach to maintaining biological diversity. *Conservation Biology* 2:375–385.

Hussey, B. M. J., R. J. Hobbs, and D. A. Saunders. 1989. *Guidelines for bush corridors*. Western Australia: CSIRO Division of Wildlife and Ecology, Western Australian Laboratory.

Ims, R. A., and N. G. Yoccoz. 1997. Studying transfer processes in metapopulations: Emigration, migration and colonization. In I. Hanski and M. Gilpin (eds.), *Metapopulation Biology: Ecology, genetics, and evolution*. San Diego: Academic Press.

IUCN. 1980. *World conservation strategy*. Gland, Switzerland: IUCN.

Janzen, D. H. 1979. How to be a fig. *Annual Review of Ecology and Systematics* 10:13–51.

———. 1983. Insects. In D. H. Janzen (ed.), *Costa Rican natural history*. Chicago: University of Chicago Press.

———. 1986. The eternal external threat. In M. E. Soulé (ed.), *Conservation biology. The science of scarcity and diversity*. Northampton, Mass.: Sinauer.

———. 1989. The evolutionary biology of national parks. *Conservation Biology* 3:109–112.

Junk, W. 1989. Flood tolerance and tree distribution in central Amazonian floodplains. In L. B. Holm-Nielsen, I. C. Nielson, and H. Balsley (eds.), *Tropical forests: Botanical dynamics, speciation and diversity*. Orlando, Fla.: Academic Press.

Koenig, W. D., D. V. Duren, and P. N. Hooge. 1996. Detectability, philopatry, and the distribution of dispersal distances in vertebrates. *Trends in Ecology and Evolution* 11:514–517.

Kohn, M. H., and R. K. Wayne. 1997. Facts from feces revisited. *Trends in Ecology and Evolution* 12:223–227.

Klein, B. C. 1989. The effects of forest fragmentation on dung and carrion beetle (Scarabaeinae) communities in central Amazonia. *Ecology* 70:1715–1725.

Kruess, A., and T. Tscharntke. 1994. Habitat fragmentation, species loss, and biological control. *Science* 264:1581–1584.

Lamb, D., J. Parrotta, R. Keenan, and N. Tucker. 1997. Rejoining habitat remnants. In W. F. Laurance and R. O. Bierregaard (eds.), *Tropical forests remants*. Chicago: University of Chicago Press.

Lamberson, R. H., B. R. Noon, and K. S. McKelvey. 1994. Reserve design for territorial species—the effects of patch size and spacing on the viability of the Northern spotted owl. *Conservation Biology* 8:185–195.

Lande, R. 1988. Demographic models of the Northern spotted owl (*Strix occidentalis*). *Oecologia* 75:601–607.

Lande, R. 1988. Genetics and demography in biological conservation. *Science* 241:1455–1460.

Langton, T. E. S. (ed.). 1989. *Amphibians and roads*. Shefford, England: ACO Polymer Products Ltd.

Laurance, W. F., and R. O. Bierregaard (eds.). 1997. *Tropical forest remnants*. Chicago: University of Chicago Press.

Leberg, P. L. 1990. Genetic considerations in the design of introduction programs. *Transactions of the North American Wildlife and Natural Resource Conference* 55:609–619.

Leigh Jr., E. G., S. J. Wright, F. A. Herre, and F. E. Putz. 1993. The decline of tree diversity on newly isolated tropical islands: A test of a null hypothesis and some implications. *Evolutionary Ecology* 7:76–102.

Little, C. E. 1990. *Greenways for America*. Baltimore: Johns Hopkins University Press.

Liu, J. G., J. B. Dunning, and H. R. Pullium. 1995. Potential effects of a forest management plant on Bachman's sparrows (*Aimophila acstivalis*) linking a spatially explicit model with GIS. *Conservation Biology* 9:62–75.

MacArthur, R. H., and E. O. Wilson. 1967. *The theory of island biogeography*. Princeton, N.J.: Princeton University Press.

McCallum, H., and A. Dobson. 1995. Detecting disease and parasite threats to endangered species. *Trends in Ecology and Evolution* 10:190–194.

McCullough, D. R. 1996. *Metapopulations and wildlife conservation*. Washington, D.C.: Island Press.

McIntyre, S., and G. W. Barrett. 1992. Habitat variegation: An alternative to fragmentation. *Conservation Biology* 6:146–147.

Meffe, G. K., and C. R. Carroll. 1994. *Principles of conservation biology*. Sunderland, Mass.: Sinauer.

Merriam, G. 1991. Corridors and connectivity: Animal populations in heterogeneous environments. In D. A. Saunders and R. J. Hobbs (eds.), *Nature conservation 2: The role of corridors*. Chipping Norton, New South Wales: Surrey Beatty & Sons.

Mills, L. S., and P. E. Smouse. 1994. Demographic consequences of inbreeding in remnant populations. *American Naturalist* 144:412–431.

Mills, L. S. 1995. Edge effects and isolation: Red-backed voles on forest remnants. *Conservation Biology* 9:395–403.

Mills, L. S., and F. W. Allendorf. 1996. The one-migrant-per-generation rule in conservation and management. *Conservation Biology* 10:1509–1518.

Mitton, J. 1994. Molecular approaches to population biology. *Annual Review of Ecology and Systematics* 25:45–69.

Moore, N. W., M. D. Hooper, and B. N. K. Davis. 1967. Hedges. I. Introduction and reconnaissance studies. *Journal of Applied Ecology* 4:201–220.

Moore, F. R., and T. R. Simons. 1992. Habitat suitability and stopover ecology of neotropical landbird migrants. In J. M. Hagan III and D. W. Johnston (eds.), *Ecology and conservation of neotropical migrant landbirds*. Washington, D.C.: Smithsonian Institution Press.

Morin, P. A., and D. S. Woodruff. 1996. Noninvasive genotyping for vertebrate conservation. In T. B. Smith and R. K. Wayne (eds.), *Molecular genetic approaches in conservation*. New York: Oxford University Press.

Moritz, C., T. E. Dowling, and W. M. Brown. 1987. Evolution of animal mitochondrial DNA: relevance for population biology and systematics. *Annual Review of Ecology and Systematics* 18:269–292.

Morris, W. F. 1993. Predicting the consequences of plant spacing and biased movement for pollen dispersal by honey bees. *Ecology* 74:493–500.

Mundy, N. I., C. S. Winchell, T. Burr, and D. S. Woodroff. 1997. Microsatellite variation and microevolution in the critically endangered San Clemente Island loggerhead shrike. *Proceedings of the National Academy of Sciences* 264:869–875.

Myers, J. P., R. I. G. Morrison, P. Z. Antas, B. A. Harrington, T. E. Lovejoy, M. Sallaberry, S. E. Senner, and A. Tarak. 1987. Conservation strategy for migratory species. *American Scientist* 75:18–26.

Nabhan, G. P., and T. H. Fleming. 1993. Endangered mutualisms. *Conservation Biology* 7:457–459.

Naiman, R. J., H. Decamps, and M. Pollock. 1993. The role of riparian corridors in maintaining regional biodiversity. *Ecological Applications* 3:209–212.

Newman, D. K. 1996. *Importance of genetics on survival of small populations: Genetic drift, inbreeding, and migration*. Missoula, Mont.: University of Montana.

Newman, D., and D. Pilson. 1997. Increased probability of extinction due to decreased genetic effective population size: Experimental populations of *Clarkia pulchella*. *Evolution* 51:354–362.

Nicholls, A. O., and C. R. Margules. 1991. The design of studies to demonstrate the biological importance of corridors. In D. A. Saunders and R. J. Hobbs (eds.), *Nature conservation 2: The role of corridors*. Chipping Norton, New South Wales: Surrey Beatty & Sons.

Nicholson, A. J., and Bailey, V. A. 1935. The balance of animal populations. Part I. *Proceedings of the Zoological Society of London:* 551–598.

Noss, R. F. 1991. Landscape connectivity: Different functions at different scales. In W. E. Hudson (ed.), *Landscape linkages and biodiversity*. Washington, D.C.: Island Press.

Okubo, A. 1980. *Diffusion and ecological problems: Mathematical models*. New York: Springer.

Olivieri, I., D. Couvet, and P.-H. Gouyon. 1990. The genetics of transient populations: Research at the metapopulation level. *Trends in Ecology and Evolution* 5:207–210.

Olivieri, I., Y. Michalakis, and P.-H. Gouyon. 1995. Metapopulation genetics and the evolution of dispersal. *American Naturalist* 146:202–228.

Paetkau, D., W. Calvert, I. Stirling, and C. Strobeck. 1995. Microsatellite analysis of population structure in Canadian polar bears. *Molecular Ecology* 4:347–354.

Paetkau, D., L. P. Waits, P. L. Clarkson, L. Craighead, E. Vyes, R. Ward, and C. Strobeck. 1998. Variation in genetic diversity across the range of North American brown bears. *Conservation Biology* 12:418–429.

Paton, P. W. C. 1994. The effect of edge on avian nest success: How strong is the evidence? *Conservation Biology* 8:17–26.

Peters, R. H. 1983. *The ecological implications of body size.* Cambridge: Cambridge University Press.

Peters, R. L. 1988. The effect of global climatic change on natural communities. In E. O. Wilson (ed.), *Biodiversity.* Washington, D.C.: National Academy Press.

Pollard, E., M. D. Hooper, and N. W. Moore. 1974. *Hedges.* London: Collins.

Powell, G. V. N., and R. Bjork. 1995. Implications of intratropical migration on reserve design: A case study using *Pharomachrus mocinno. Conservation Biology* 9:354–362.

Pullium, H. R. 1988. Sources, sinks, and population regulation. *American Naturalist* 132:652–661.

Pullium, H. R., J. B. Dunning, and J. G. Liu. 1992. Population dynamics in complex landscapes—a case study. *Ecological Applications* 2:165–177.

Ramirez, B. W. 1970. Host specificity of fig wasps (Agaonidae). *Evolution* 24:681–691.

Rannala, B., and J. A. Hartigan. 1996. Estimating gene flow in island populations. *Genetical Research* 67:147–158.

Roca, R. L. 1994. Oilbirds of Venezuela: Ecology and conservation. *Publications of the Nuttall Ornithological Club.* Cambridge, Mass. 83 pp.

Rosenberg, D. K., B. R. Noon, and C. Meslow. 1997. Biological corridors: Form, function and efficacy. *BioScience* 47:677–687.

Roy, M. S., E. Geffen, D. Smith, E. A. Ostrander, and R. K. Wayne. 1994. Patterns of differentiation and hybridization in North American wolflike canids, revealed by analysis of microsatellite loci. *Molecular Biology Evolution* 11:553–570.

Ruckelshaus, M., C. Hartway, and P. Kareiva. 1997. Assessing the data requirements of spatially explicit dispersal models. *Conservation Biology* 11:1298–1306.

Russell, R. W., F. L. Carpenter, M. A. Hixon, and D. C. Paton. 1994. The impact of variation in stopover habitat quality on migrant rufous hummingbirds. *Conservation Biology* 8:483–490.

Saunders, D. A., and R. J. Hobbs. 1991. *Nature conservation 2: The role of corridors.* Chipping Norton, New South Wales: Surrey Beatty & Sons.

Schoener, T. W. 1983. Rate of species turnover decreases from lower to higher organisms: A review of the data. *Oikos* 41:372–377.

———. 1968. Sizes of feeding territories among birds. *Ecology* 49:123–141.

Schumaker, N. H. 1996. Using landscape indices to predict habitat connectivity. *Ecology* 77:1210–1225.

———. 1998. *A user's guide to the PATCH model*. Corvallis, Ore.: U.S. Environmental Protection Agency.

Schwartz, M., D. Tallmon, and G. Luikart. 1998. Review of DNA-based census and effective population size estimators. *Animal Conservation* 1:293–299.

Simberloff, D. 1974. Equilibrium theory of island biogeography and ecology. *Annual Review of Ecology and Systematics* 5:161–182.

———. 1988. The contribution of population and community biology to conservation science. *Annual Review of Ecology and Systematics* 19:473–511.

———. 1993. Species-area and fragmentation effects on old-growth forests: Prospects for longleaf pine communities. In S. M. Hermann (ed.), *Proceedings of the tall timbers fire ecology conference: The longleaf pine ecosystem—ecology, restoration, and management*. Tallahassee, Fla.: Tall Timbers Research Station.

Simberloff, D., and J. Cox. 1987. Consequences and costs of conservation corridors. *Conservation Biology* 1:63–71.

Simberloff, D., J. A. Farr, J. Cox, D. W. Mehlman. 1992. Movement corridors: Conservation bargains or poor investments? *Conservation Biology* 6:493–504.

Sjogren, P., and Wyoni, P. 1994. Conservation genetics and detection of rare alleles in finite populations. *Conservation Biology* 8:267–275.

Skellam, J. G. 1951. Random dispersal in theoretical environments. *Biometrika* 38:196–218.

Slatkin, M. 1985. Gene flow in natural populations. *Annual Review of Ecology and Systematics* 16:393–430.

———. 1994. Gene flow and population structure. In L. A. Real (ed.), *Ecological genetics*. Princeton, N.J.: Princeton University Press.

———. 1995. A measure of population subdivision based on microsatellite allele frequencies. *Genetics* 139:457–462.

Slatkin, M., and N. H. Barton. 1989. A comparison of three indirect methods for estimating average levels of gene flow. *Evolution* 43:1349–1369.

Soulé, M. E. 1983. What do we really know about extinction? In S. M. Schonewald-Cox et al. (eds.), *Genetics and Conservation*. Reading, Mass.: Addison Wesley.

———. 1991. Theory and strategy. In W. E. Hudson (ed.), *Landscape linkages and biodiversity*. Washington, D. C.: Island Press.

Soulé, M. E., and D. Simberloff. 1986. What do genetics and ecology tell us about the design of nature refuges? *Biological Conservation* 35:19–40.

Soulé, M. E., and M. E. Gilpin. 1991. The theory of wildlife corridor capability. In D. A. Saunders and R. J. Hobbs (eds.), *Nature conservation 2: The role of corridors*. Chipping Norton, New South Wales: Surrey Beatty & Sons.

Soulé, M. E., and M. Sanjayan. 1998. Conservation targets: Do they help? *Science* 279: 2060–2061.

Spielman, D., and R. Frankham. 1992. Modeling problems in conservation genetics using captive Drosophila populations: Improvement of reproductive fitness due to immigration of one individual into small partially inbred populations. *Zoo Biology* 11:343–351.

Taberlet, P., J. J. Camarra, S. Griffin, E. Uhres, O. Hanotte, L. P. Waits, C. Dubois-Paganon, T. Burke, and J. Bouvet. 1997. Noninvasive genetic tracking of the endangered Pyrenean brown bear population. *Molecular Ecology* 6:869–876.

Taylor, P. D., L. Fahrig, K. Henein, and G. Merriam. 1993. Connectivity is a vital element of landscape structure. *OIKOS* 68(3):571–573.

Templeton, A. R., and N. J. Georgiadis. 1995. A landscape approach to conservation genetics: Conserving evolutionary processes in the African Bovidae. In J. Avise and J. Hamrick (eds.), *Conservation genetics: Case histories from nature.* New York: Chapman & Hall.

Templeton, A. R., E. Routman, and C. A. Phillips. 1995. Separating population structure from population history: A cladistic analysis of the geographical distribution of mitochondrial DNA haplotypes in the tiger salamander (*Ambystoma tigrinum*). *Genetics* 140:767–782.

Terborgh, J. 1974. Preservation of natural diversity: The problem of extinction prone species. *BioScience* 24:715–722.

Terborgh, J., and B. Winter. 1980. Some causes of extinction. In M. E. Soulé and B. A. Wilcox (eds.), *Conservation biology: An evolutionary-ecological perspective.* Sunderland, Mass.: Sinauer.

Thomas, C. D. 1994. Extinction, colonization, and metapopulations: Environmental tracking by rare species. *Conservation Biology* 8:373–378.

Turchin, P. 1998. *Quantitative analysis of movement: Measuring and modeling population redistribution in animals and plants.* Sunderland, Mass.: Sinauer.

Turner II, B. L., W. C. Clark, R. W. Kates, J. F. Richards, J. T. Mathews, and W. B. Meyer. 1990. *The Earth as transformed by human action.* Cambridge: Cambridge University Press.

Turner, F. B., R. I. Jennrich, and J. D. Weintraub. 1969. Home range and body size of lizards. *Ecology* 50:1076–1081.

Turner, M. G., F. H. Sklar, and R. Costanza. 1989. Methods to evaluate the performance of spatial simulation-models. *Ecological Modelling* 48:1–2, 1–18.

Valiente-Banuet, A., M. D. C. Arizmendi, A. Rojas-Martinez, and L. Dominguez. 1996. Ecological relationships between columnar cacti and nectar-feeding bats in Mexico. *Journal of Tropical Ecology* 12:103–119.

van den Bosch, F., R. Hegenfeld, and J. A. J. Metz. 1992. Analysing the velocity of animal range expansion. *Journal of Biogeography* 19:135–150.

Varvio, S., R. Chakraborty, and M. Nei. 1986. Genetic variation in subdivided populations and conservation genetics. *Heredity* 57:189–198.

Wahlberg, N., A. Moilanen, and I. Hanski. 1996. Predicting the occurrence of endangered species in fragmented landscapes. *Science* 273:1563–1538.

Warkentin, I. G., R. Greenberg, and J. S. Ortiz. 1995. Songbird use of gallery woodlands in recently cleared and older settled landscapes of the Selva Lacandona, Chiapas, Mexico. *Conservation Biology* 9:1095–1106.

Waser, P. M., and C. Strobeck. 1998. Genetic signatures of interpopulation dispersal. *TREE (Trends in Ecology and Evolution)* 13:43–44.

Wegner, J. F., and G. Merriam. 1979. Movements by birds and small mammals between a wood and adjoining farmland habitats. *Journal of Applied Ecology* 16:349–357.

Wilcove, D. S., C. H. McLellan, and A. P. Dobson. 1986. Habitat fragmentation in the temperate zone. In M. E. Soulé (ed.), *Conservation biology.* Sunderland, Mass.: Sinauer.

Wilkinson, G. S., and T. H. Fleming. 1996. Migration and evolution of lesser long-nosed bats (*Leptonycteris curasoae*) inferred from mitochondrial DNA. *Molecular Ecology* 5:329–339.

Willis, E. O. 1984. Conservation, subdivision of reserves, and the antidismemberment hypothesis. *OIKOS* 42:396–398.

Wilson, M. H., C. B. Kepler, N. R. F. Snyder, S. R. Derrickson, F. J. Dein, J. W. Wiley, J. M. Wunderle Jr., A. E. Lugo, D. L. Graham, and W. D. Toone. 1994. Puerto Rican parrots and potential limitations of the metapopulation approach to species conservation. *Conservation Biology* 8:114–123.

Winker, K., D. W. Warner, and A. R. Weisbrod. 1992. The northern waterthrush and Swainson's thrush as transients at a temperate inland stopover site. In J. M. Hagan III and D. W. Johnston (eds.), *Ecology and conservation for neotropical migrant landbirds*. Washington, D.C.: Smithsonian Institution Press.

With, K. A. 1997. The application of neutral landscape models in conservation biology. *Conservation Biology* 11:1069–1080.

Woodroffe, R., and J. R. Ginsberg. 1998. Edge effects and the extinction of populations inside protected areas. *Science* 280:2126–2128.

Woolhouse, M. E. J. 1998. Patterns in parasite epidemiology: The peak shift. *Parasitology Today* 14:428–434.

Wright, S. 1931. Evolution in Mendelian populations. *Genetics* 16:97–259.

Wright, S. 1932. The roles of mutation, inbreeding, crossbreeding, and selection in evolution. *Proceedings of the Sixth International Congress of Genetics* 1:356–366.

# 7 Buffer Zones: Benefits and Dangers of Compatible Stewardship

*Martha Groom, Deborah B. Jensen, Richard L. Knight, Steve Gatewood, Lisa Mills, Diane Boyd-Heger, L. Scott Mills, and Michael E. Soulé*

Totally protected nature reserves comprise about 2 percent of the land area in Mexico, 4 percent in Canada, and 7.5 percent in the United States. Studies, however, indicate that it may take 35 to 75 percent of the land surface to prevent significant reductions to species diversity (Soulé and Sanjayan 1998). The majority of rare and endangered species in the United States do not have populations within nature preserves. Nor are all the continent's ecosystems well represented in such areas (Crumpacker et al. 1988; Hummel 1989). For this reason, conserving biodiversity in North America will depend, to a considerable degree, on what happens on lands outside existing reserves. To this end, The Nature Conservancy is developing "portfolios" of reserves in all ecoregions in the United States, and The Wildlands Project is planning an extensive network of wildland conservation areas in North America consisting of strictly protected and interconnected core areas. Even so, unreserved lands will always contain much, if not most, of the species and ecosystems in jeopardy.

Consider California. About 15 percent of the land area of California had been converted entirely to urban or agricultural uses by the late 1980s, and about 12 percent had been designated as reserves (with roughly half of these areas under multiple-use designation rather than strict protection; Jensen et al. 1993). The remaining 72 percent of the state's land area, most of it mountain and desert, had multiple land uses. While California has hundreds of species at risk and has lost populations of many of the charismatic animals such as tule elk (*Cervus elephas*) and brown (grizzly) bear (*Ursus arctos*), most of the species still persist in viable populations amid the extensive unprotected, yet not utterly denatured, areas of the state. Politically aware biologists know that biodiversity conservation requires a mix of strict nature reserves and sound management of the seminatural matrix (Harris 1984; Brown 1988; Franklin 1993). Because such a small percentage of land is protected, designers must plan for compatible activities in many of these matrix areas. The type and intensity of human uses in these buffer zones or "surrounds" must be arranged, and managers must adopt practices that permit the persistence of biodiversity. The design and management of buffers are rarely addressed, however, because analysis involves a complex suite of variables and there is a greater likelihood of land-use conflicts in buffers than in dedicated core areas.

This chapter focuses on the knotty issues of when human uses are compatible with conservation goals, explains how conflicts and compatibility may change over time, and explores incentives that encourage compatible forms of development within such areas. We conclude with some practical advice on how to incorporate compatible stewardship into reserve networks and point out some opportunities and challenges for the future.

## What Are Buffers?

Buffers are areas that maintain some degree of wildness but allow sustainable economic uses that are compatible with the goals of the reserve network as a whole. Essentially buffers insulate the core areas and connecting corridors from deleterious influences. Buffers differ from core areas in the degree to which they serve as natural areas and as regions of human economic activity. Shelford (1933) first proposed buffers to protect nature sanctuaries. Forty years later, the buffer concept was adopted by the biosphere reserve program to distinguish a transition zone that surrounds and protects natural core areas and primary corridors (UNESCO 1974). Since then the concept has been most developed in the tropics, though without much critical scrutiny. Recently Noss and Cooperrider (1994) defined the buffer zone in the context of reserve networks as "an area that permits a greater range of human uses than core reserves, but still is managed with native biodiversity as the preeminent concern."

Depending on the area, a buffer zone may maintain virtually all or only a subset of the ecosystem processes and focal species typical of core areas. At one extreme, buffers could do more harm than good for native species. Buffers can be demographic sinks for species sensitive to human activity, for example, or subject to poaching or roadkill, sentencing individuals within them to an almost certain and early death or at least to reproductive failure. At the other extreme, species may flourish in buffer zones. In the middle are species that can tolerate certain intensive human activity, particularly in areas they visit infrequently. In practice this means that the buffer's effectiveness depends on the species in question, the ecological, human demographic, and economic uses of the area, and how they change over time.

The ambiguity of designating lands as "core" or "buffer" is evident. For some species, designated or potential buffer zones function completely as core habitat; for many others, however, buffers provide supplemental habitat that increases the size or viability of the population. Species that can be maintained entirely in buffer zones include many plant species, insects, amphibians, and other animals with small and restricted ranges. For most species, though, buffer areas could serve only as supplemental habitat. As part of the Safe Harbor Program of the U.S. Fish and Wildlife Service, for example, the network of small parcels of land used for wood and pine duff production in the sandhills region of North Carolina supplements suitable habitat for red-cockaded woodpeckers (*Picoides borealis*) on larger parcels of protected lands. Although these exploited landholdings are not sufficient to maintain viable populations of red-cockaded woodpeckers, they contribute to the stability of populations.

For species requiring large areas, buffer zones may be a critical component of a network of sites that together constitute the essential range for a population. While Yellowstone National Park itself is not large enough to sustain independent, viable populations of wolves, for example, the surrounding network of multiuse lands, if properly connected and managed, may suffice. Similarly, privately owned farms and ranchlands in the western United States often serve as winter range for elk, and some ranches agree to provide year-round habitat for these ungulates.

Buffer areas are dynamic and complex entities in terms of their function in a reserve network. Not only do buffer zones facilitate connectivity between reserves, but occasionally they may function as cores or as refugia by providing the resources of a reserve if disturbances make reserve habitats temporarily unsuitable (Noss and Cooperrider 1994). Depending on the species, the community, and the activities in and around the buffer zone, therefore, buffers can provide many of the ecological functions of cores and corridors. Planners and managers must remain flexible about these labels and uses, particularly as land uses and climate change.

Some critics have suggested that the term "buffer" is negative because it implies that civilization is at war with nature, that nature needs protection from people, or that human beings are not a healthy part of nature. Alternative constructions such as "zones of compatible stewardship" may convey more information and imply a greater potential for harmony between society and nature, but they are too cumbersome for frequent use. For simplicity, however, we will use the term "buffer."

## Defining Compatible Uses

For a buffer zone to have conservation value, land uses must be compatible with the goals of conservation in the region. Compatible uses are those that do not interrupt processes such as nutrient cycling, key interactions among species, or reproduction. Incompatible uses are those that interrupt processes, compromise ecosystem integrity or population viability, or cause irreversible changes in these processes. Obviously, the degree of compatibility of human activities with wildland conservation differs widely among uses (such as nonmotorized outdoor recreation versus motorized recreation) and depends on the scale of activity. As the number of outdoor enthusiasts grows, for example, recreational use may threaten conservation values due to the added noise, trampling, pollution, and erosion (Petersen 1996). Large ranches in the arid West may have considerable conservation value due to their extensive areas of undisturbed land, yet this compatibility may disappear if the ranches are sold and developed into suburban residential areas (Knight et al. 1995). Further, a use that is compatible in one area may be incompatible in another. Finally, the cumulative impacts of human activity may determine species' responses.

For nearly all human activities there is a gradient between compatible and incompatible uses. This gradient is created by differences in the intensity of use (or the number or density of people engaged in an activity), the duration of the activity, the frequency of use, the timing of that use relative to annual or superannual cycles in the environment, and life histories of the organisms in the environment. Gradients of human use may have nonlinear or threshold effects on species or ecosystems (Figure 7.1). Moreover, because the effects of many human activities are cumulative, they should not be considered independent.

To illustrate these complexities, we now turn to several examples of

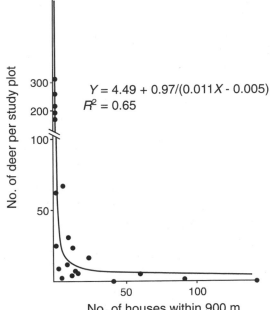

Figure 7.1. Curvilinear relationship between mule deer densities and densities of rural homes in Montana. Adapted from Vogel (1989).

common human activities in buffer zones. These examples demonstrate how changes in the intensity, duration, or extent of activity tend to move these uses along a continuum from "compatible" to "incompatible" with the goals of conservation. For a summary and additional examples see Table 7.1.

## Agriculture

A variety of bird and mammal species are able to persist in forest or grassland fragments embedded in an agricultural matrix—provided fragments are sufficiently large, numerous, or interconnected (Andren and Angelstam 1988). For example, Heske (1995) found no differences in indices of abundance in forest interior versus agricultural edge for mammal species. Many migrating wading birds or waterfowl benefit from feeding in fields during winter. In California's Central Valley, more than 90 percent of the original wetlands have been drained and developed.

## Table 7.1. Uses and degree of compatibility within buffer areas.

| | Use and Effects |
|---|---|
| Outdoor recreation | Zone of influence determined by type of activities (motorized or nonmotorized), noise level, whether on or off trail or road, day use only or camping overnight, during breeding season or not, during peak foraging seasons or not, density, frequency and duration of activities allowed in conjunction, pollution, and level of harassment (unintentional and intentional), soil compaction or alteration of plant communities. |
| Ecotourism | Zone of influence determined by size of group, number of groups, time of year groups visit, length of stay, type of activities group engages in, and size of area that group visits (and Outdoor Recreation list). |
| Residential development | Zone of influence determined by density of homes, size of homes, seasonality of occupancy, whether clustered or dispersed, road traffic, whether pets are kenneled or allowed to roam free, whether landscaping is with native or exotic species, whether outside lights are left on or not, the degree of human use of the surrounding areas, design and placement of roads and fences. |
| Mining activity | Zone of influence determined by size of operation, duration of operation (seasonality and number of years), access roads and level of traffic, level of noise in conjunction with operation, use of toxics and hazardous materials, containment of contaminants, potential for spills, degree of restoration following operation, and education of personnel involved in operation. |
| Logging activities | Zone of influence determined by type of logging activity, size of logging operations, use of motorized machinery, time of year, number, location, and type of road construction, silt containment, management of personnel involved in operation, recreational activities of personnel, and post-logging restoration. |

| Livestock grazing | Zone of influence determined by number and type of animals grazed, areas where grazing is allowed, rotational practices, time of year and duration of grazing, interval of rest allowed between grazing, fencing of streams and riparian areas, and type of access allowed to put livestock on the area. |
|---|---|
| Farming | Zone of influence determined by intensity, tolerance of woodlots, persistence of unplowed strips and fence rows, presence of shelter belts, tolerance of undrained wetlands, use of pesticides, types of fertilizers, source of irrigation water, and recharge of aquifer. |
| Transportation facilities | Zone of influence determined by kinds of roads, traffic levels, nighttime driving, wildlife overpasses and underpasses, setbacks for trees and mowing along verges, intensity of rail traffic, airport traffic, and size of airports and other transportation hubs. |
| Towns, resorts | Zone of influence determined by degree of sprawl, kind of planning process, zoning, tolerance of strip development, sewage treatment, water sources, lighting regulations, pet regulations and control, and seasonality of use. |

The Cosumnes River Preserve is an effort to restore and maintain a number of the communities and species characteristic of the Sacramento Valley. Here valley oak woodlands and the naturally flooding river systems are maintained. The reserve is bordered in some places by organic rice growers who farm on lands that were once intermittently flooded grasslands. The farmers now enjoy the presence of overwintering sandhill cranes (*Grus canadensis*) in their fields. Although these farmlands are too intensively used to be considered core for migrating sandhill cranes, the farms might become a critical component of a reserve network because they would help increase the food resources available in winter.

When the agricultural matrix is too extensive, however, bird species may suffer lower reproductive success and decline—for example, as prairie habitats shrink to linear habitats at the margins of fields (Warner 1994). Agricultural activities may conflict with conservation in buffer zones through the effects of pesticides, fertilizers, and water runoff or erosion. In New South Wales, for example, native plant species richness declined with increases in soil fertility and water runoff from agriculture and livestock grazing (McIntyre and Lavorel 1994). As agricultural activities rely more heavily on fertilizers, pesticides, or scarce water supplies, they will come into greater conflict with wildlands conservation—even if the extent of agricultural fields is not excessive. Predation on bird nests at remnant edges may increase as the contrast between the remnants and buffer increases (Angelstam 1986; Ratti and Reese 1988; Robinson et al. 1995); similar trends have been documented for mammals (Lidicker 1995).

## Logging

A key variable governing the compatibility of logging in buffer areas is the degree of contrast between forest and buffer. When part or all of a buffer is clear-cut, species distributions on fragments are almost certain to change. In tropical forests of Queensland, for example, Laurance (1994, 1995) found that the ability of a species to tolerate the surrounding matrix was the most important predictor of vulnerability—probably because species that used the matrix could easily move between forest remnants and were preadapted to subsequent changes in remnants. Differences in vulnerability led to changes in species composition on the remnants. These patterns are also seen extensively in populations of breeding birds that require forested habitats.

The importance of the contrast between remnant and buffer has been demonstrated in the Biological Dynamics of Forest Fragments Project in Manaus, Brazil. Here bird, primate, coprophagous beetle, and leaf-litter beetle communities were less abundant and diverse in fragments surrounded by regenerating secondary forest than in continuous forest—implying isolation by the matrix (Bierregaard et al. 1992; Klein 1989; Bierregaard and Stouffer 1997; Didham 1997)—but some mammals increased in number of species and abundance on the fragments (Malcolm 1997). When the fragments were surrounded by a harsher matrix such as pasture, however, all species including the mammals were negatively affected (Malcolm 1997). Similarly, for some breeding birds, forest fragmentation due to logging may be less harmful than fragmentation due to agriculture (Bayne and Hobson 1997)—although in this case the difference is less attributable to changes in microclimate than to increased predation as predator species increase in density in agricultural areas.

In the Pacific Northwest of the United States, small-mammal studies demonstrate how the degree and type of logging can affect the quality of the buffer for ecological interactions. California red-backed voles (*Clethrionomys occidentalis*) suffer from negative edge effects and from isolation on forest remnants surrounded by three- to thirty-year-old clearcuts in southwestern Oregon—probably because of the loss of their preferred food, belowground mycorrhizal fungi sporocarps (Clarkson and Mills 1994; Rosenberg et al. 1994; Mills 1995, 1997). Other species show no changes in abundance on the remnants, while deer mice (*Peromyscus* spp.) are positively affected (Mills 1997). The success of deer mice on both burned and unburned clearcuts, and at the edges of forest fragments, appears to be a general trend in coniferous forests (Gashwiler 1970; Kirkland 1990; Sullivan 1979; Martell 1983). Some of the consequences of deer mouse abundance can be deleterious, however. Jules et al. (in press) studied the understory herb *Trillium ovatum* in the same areas studied by Mills (1995, 1997) and found that recruitment is near zero—both on clearcuts and within 65 meters of forest-clearcut edges—in large part due to seed predation by the abundant deer mice. Moreover, Jules et al. (in press) found decreased pollination by beetles and bees outside of forest fragments. See Chapter 3 for similar examples from the tropics.

The compatibility of logging with conservation goals may depend on the methods used to cut and remove trees. The degree of soil disruption, compaction, and loss, the extent or intensity of the cut, whether roads are reconfigured and closed, the investment in and success of future management—all are variables that determine the degree of compatibility. For example, three different silvicultural management techniques (the use of fire and two different site-preparation techniques following clear cutting) resulted in different reptile assemblages in a central Florida pine scrub habitat (Greenberg et al. 1994). Selectively logged forests and plantations may be compatible with the conservation of a large number of species, although such forests rarely harbor as diverse a fauna or flora as natural forest (Johns 1985; Raivio and Haila 1990; Thiollay 1995).

Road density is another critical variable in determining the compatibility of logging with wildlife conservation. For species harmed by roads and by the presence of humans, such as grizzly, black bear (*Ursus americanus*), and wolverine (*Gulo gulo*), even light logging activities may make buffer areas dangerous—not only because of the logging per se but because the logging roads facilitate legal and illegal access by humans long after the logging operation is completed. Most species will begin to be adversely affected as the density of roads increases. A threshold is thought to exist at about 0.6 kilometer of road per square kilometer; above this density, some critical or sensitive species decline severely (Cole et al. 1997; Forman et al. 1997; Baker and Knight 1999).

## Grazing

One of the most contentious land-use issues is grazing of public and private lands over large areas of the West, particularly the more arid regions. For this use, the complete gradient between compatible and incompatible grazing practices can readily be found. Cattle often degrade riparian areas, causing severe erosion and reducing plant and animal diversity (Dobkin et al. 1998). In some situations, however, grazing in buffer areas has conservation benefits. The presence of cattle in Palo Verde National Park in Costa Rica, for example, may suppress the growth of nonindigenous cattails in the marshes of the park, benefiting waterfowl and some other species (McCoy 1994). But it is critical that cattle are neither so numerous that their wastes substantially change nutrient loads in the marsh nor graze in other areas of the park on native species. Thus cattle ranching is a beneficial practice in this national park so long as it is carried out at a suitable intensity and managed with regard to the native vegetation in the park. In practice, however, it is difficult indeed to regulate grazing.

Similarly, Colwell and Dodd (1995) document that cattle ranching in coastal habitats in northern California can benefit wildlife if grazing is managed in such a way as to maintain a mosaic of vegetation heights, and to open watercourses. Defining the appropriate intensity of grazing for a site—and regulating it—will determine whether or not grazing is a compatible use in a buffer zone. Grazing in the absence of scrutiny and enforcement, however, is likely to cause more harm than good.

## Residential Development

Residential development may be compatible with conservation of wildlands—so long as the development is highly concentrated or remains at a very low level and does not lead to fragmentation of remaining habitats. A growing body of literature reveals negative relationships between density of housing and the health of wildlife populations. Neotropical migrant birds sharply decline in abundance when houses are built at the margins of forest patches, for example, and overall show strong negative correlations in abundance with housing density (Friesen et al. 1995; Blair 1996; Knight and Clark 1998; Clergeau et al. 1998). Yet residential development is increasing throughout the United States—especially in traditionally ranching and farming areas of the West (Knight 1997; see Figure 7.2). In the eastern United States, it is quite difficult to find large tracts devoid of human settlement, and extensive regions of rural character may be the only options for buffers.

Human activity adjacent to and within public lands certainly reduces the utility of these areas to buffer wildlife populations in cores. Increased densities of buildings and roads can alter native species diversity on both sides of the public/private land border (Knight and Clark 1998). Some of the proximal factors include increases in domestic animals that prey on wildlife, increased automobile traffic, and night lights (Knight et al. 1995). Subdivisions in the arid West can cause further harm by depleting water resources. Domestic animals, particularly house cats, appear to constitute a threat because they occur at densities many times greater than native predators on birds, reptiles, and small mammals (Soulé et al. 1988; Crooks 1997). Human-adapted species, both native and alien, tend to

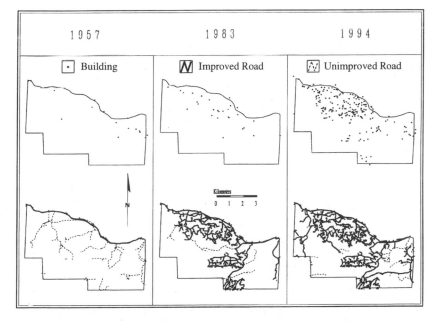

Figure 7.2. Increase in the density of rural homes in Larimer County, Colorado. Adapted from Baker and Knight (1999).

gain as housing densities increase—all at the expense of sensitive, native species (Knight 1997).

Another consequence of increased human presence in and around protected areas is more conflict for the managing agencies. Bears, deer, and elk may cross from public lands onto private property, for example, creating a "nuisance" (Knight and Clark 1998). Forbes and Theberge (1996) document how Algonquin Park in Ontario serves as a population source for gray wolves (*Canis lupus*), yet these animals frequently leave the park to hunt on surrounding private lands. Private landowners feel that wolves encroach on their rights and threaten their safety. In this region, most wolf deaths attributed to humans occurred across the park's borders on private lands. In another example, public lands in California serve as sources for mountain lions (*Felis concolor*). With increasing housing development adjacent to these public lands, there has been a sharp increase in lion depredation on dogs and cats, an increase in lion attacks on people, and a marked increase in permits issued to kill offending lions (Torres et al. 1996).

Similarly, increased human densities and economic activities along public land borders have altered the ability of agencies to implement management activities that may cross over onto private lands (Clark and Minta 1994). For example, a more natural fire regime is being restored to public lands to recreate historic patterns of landscape heterogeneity. Fire, however, may threaten the ever-increasing number of homes rimming the urban/wildland interface (Cortner et al. 1990; Stanton 1995). Increasingly, the hands of land managers are tied when they attempt to repair the damage of decades of mismanagement.

Because the trend of increased housing and commercial development in areas that historically served as buffers for natural areas is expected to increase (Knight 1997), agencies, organizations, and individuals must plan for development that minimizes harmful ecological effects, including the clustering of housing. If private lands in regions of population growth are to serve as effective buffer zones, a period of agonizingly honest dialogue must ensue—and quickly. Without it, the essential planning and land-use agreements will not materialize in time.

## Tourism and Recreation

Because tourism and outdoor recreation often require little infrastructure and rely for their success on the maintenance of wilderness, both are seen as compatible uses of buffer areas. Tourism and recreation may further benefit conservation goals by providing economic benefits to a buffer region while simultaneously educating people about the benefits of conservation (Duffus and Dearden 1990). To a large extent the truth of this view depends on the kind of use (by foot, by horse, or by motorized vehicles) and the amount of access required for this tourism (including roads and accommodations).

Outdoor recreation involving motorized activities such as all-terrain vehicles can cause extensive erosion (Griggs and Walsh 1981) and disruption to wildlife populations, particularly where these sports gain in popularity or and activities coincide with sensitive periods for wildlife. The creation of ski resorts can have many detrimental impacts, as well, including erosion due to the construction of ski trails, resorts, and access roads, habitat fragmentation, and, if slopes are seeded with exotic grasses, introduction of exotic species (Tsuyuzaki 1994). If harmful forms of recreation are allowed and these pursuits are not managed, one might quickly find ecological degradation occurring by the accumulation, synergism, or magnification of separate activities.

A number of negative impacts can arise from nature tourism and recreation: displacement of wildlife from desirable habitat, habituation of wildlife to human presence, off-trail or site use by people and their pets, introduction of exotic plants and pathogens, trampling, littering, camping, soil compaction, attraction or avoidance behavior, shooting, poaching, vandalism, accidental fires, road-killed wildlife, chemical pollution, and the nuisance effects of noise (Dustin and McAvoy 1987; Petersen 1996). Even very low levels of visitation may alter the behavior of species, causing individuals to avoid areas frequented by tourists (Groom 1992). Many of these negative impacts can be avoided, however, with education and strict regulations. Despite the problems associated with certain forms of recreation, this still is a use of buffer zones that holds a great deal of promise for being compatible with the conservation goals of a reserve network.

## Opportunities and Dangers

Buffer zones are rarely static in quality. As the intensity and type of uses change in a buffer area, plant and animal communities will change, too, improving for some and deteriorating for others. It is critical to recognize both the opportunities and the dangers inherent in nominating multiple-use lands for conservation purposes when designing regional reserve networks.

Ideally, viable wildlife and plant communities and ecological processes can be maintained, enhanced, or reestablished in buffer zones. In reality, however, as development continues to increase adjacent to sensitive protected areas, populations of native species are likely to decline, riparian and other plant communities will be degraded, and ecological processes will be altered, generally in directions that cannot be predicted. As lands in the buffer are developed and used more intensively, the primary processes that create landscape-level diversity—such as fire, disease, predation by large carnivores, and movement of wildlife—will decrease in magnitude and frequency (Knight et al. 1995). Thus ecological degradation may come about not only from increasing intensities of use but from the substitution of deleterious uses for beneficial ones. A review of land-use history and the high rates of human population growth leads to pessimism: ecological degradation is an ever-present, and often realized, danger in the kinds of lands usually designated as buffer zones (Soulé and Sanjayan 1998).

There are, however, opportunities for increasing the conservation value of buffer zones. For example, as some intensively cultivated agrcultural lands are being abandoned in the Midwest, in the San Joaquin Valley of California, and in the Southeast, there is potential for restoring these lands or integrating them with more compatible uses. By the same token, lightly used buffer zones can be improved. The construction of artificial cavities in remnant longleaf pine habitats may increase the size of red-cockaded woodpecker populations (Copeyon et al. 1991), thus increasing the conservation value of such remnants. Buffers can shift back and forth between functional classifications and provide services that vary over time and on-site conditions. Restoration (Chapter 4) can improve buffer habitat over extended periods of time (decades to centuries). What is now a buffer may be desirable as a core or corridor in the future. This is especially true where large-scale landscape disturbance regimes, such as fire and hurricanes, operate. Moreover road closures, cessation of human use, and economic incentives may all be applied to a buffer zone in order to hasten conversion to core reserve at some future date.

A dynamic view of buffer zones is perhaps best illustrated by changes in the concept of wetland buffers over the past few decades. Riparian and wetland buffers for aquatic communities have long been promoted to protect water quality and wildlife habitat (Brown et al. 1987; Schaefer and Brown 1992). Initially these buffer zones were designed as narrow strips of vegetated habitat with a primary purpose of intercepting overland flow to reduce nutrient or pollutant inputs from watersheds and to shade the water body to protect invertebrates and vertebrates developing in the streams (FLDACS 1993; Brown et al. 1987). Increasingly, however, the question of how buffer strips may function for wildlife habitat and connectivity is being addressed in wetland mitigation and conservation projects (Schaefer and Brown 1992; Harris 1989). Further, extensive upland areas adjacent to floodplains are considered important for protecting aquatic integrity, as blocks of habitat, and as movement corridors. Thus the original buffer strips along aquatic communities where intensive forestry could still be practiced (FLDACS 1993) are now used for a greater diversity of purposes, acknowledging that with some modification of design, such areas could serve species as dispersal corridors and foraging or breeding habitat. As we think about incorporating buffer zones into regional reserve networks, we need to remember that not only will buffers change in quality, but our perspectives on what they can achieve will change as well.

Just as buffers may deteriorate over the years or improve in quality, we should consider as well whether there are times when buffers should not be incorporated into a reserve design. When buffers provide critical habitat for certain species or offer economic incentives that encourage popular support of a reserve network, they will be key components of a regional conservation plan. But buffers sometimes act as population sinks—drawing individuals away from areas where their survivorship or breeding success is high to regions where fitness is compromised—thereby reducing the long-term persistence of the population (Noss and Cooperrider 1994). Moreover, buffers may harbor nonindigenous competitors, predators, or pathogens that can enter core areas. In cases where buffers increase mortality and reduce population stability, it is worth asking whether barriers between core areas and zones of intensive human use would be more appropriate. Perhaps hard boundaries—fences or other barriers—or stark contrasts between core areas and buffer areas should be incorporated into the reserve design (as suggested by Janzen 1986). Rather than, say, attempting to accommodate a housing development or intensive agriculture adjacent to a reserve boundary, it may make more sense to create an impermeable barrier to exclude intrusion into the core. Such a barrier can minimize if not prevent the entry of disease-carrying domestic animals, domesticated or feral predators such as house cats and dogs, and casual poaching of trees and animals. In the other direction, it can prevent the incursions of wild animals onto farmlands where they can destroy crops and create resentment against the conservation activity or reserve.

Buffer zones are rarely a panacea. Their success is likely to be proportional to the strength of the conservation ethic of the buffer's human inhabitants and visitors. Moreover, they require constant monitoring and law enforcement, which have always proved difficult to achieve.

# Designing Areas of Compatible Stewardship

Presently there is no systematic approach to designing buffer areas. The characteristics of each buffer zone depend on what they are buffering, the current and probable future uses in the zone, and an institutional capacity to learn from mistakes. The flexibility required to accommodate a wide range of wilderness and human use precludes establishing a hard and fast protocol for buffer design. Nonetheless, we offer a checklist of design considerations for buffers:

## *Set Conservation Goals*

- What species, community types, or ecosystem attributes is the buffer meant to preserve for the region as a whole? A reserve network designed solely for vertebrates will differ from one designed for both plants and animals, and certainly a reserve network designed for current species will differ from one designed to accommodate all species historically present within the site or region.
- For which species is the buffer meant to provide essential habitat and resources? For what processes is it meant to be a transition zone?
- Are the political and geographic boundaries of buffers (and core areas) clearly delineated? This must be done to eliminate uncertainty.
- Do stakeholders agree on the goals of the buffer? If they do not, then disagreements about the amount and distribution of land, the focal species to be accommodated, and the compatible land uses needed in a network are likely to be perennial sources of conflict.

## *Determine Needs of Focal Species or Processes*

- What species will depend on the quality of the buffer zone?
- Is there enough good habitat for these species?
- For which species might the buffer be a sink?
- Can the focal species serve as umbrella or indicator species for other taxa of concern?
- Are certain community types required by the species?
- Is there information on how species use buffer areas or other lands with human activities?
- Is there information on the edge effects attributed to human activity

likely to occur in the buffer? (See Laurance and Gascon 1997 for operational guidelines for influencing forest fragmentation.)
- How much of the watershed needs buffer status to protect water quality?
- How might fires in the buffer affect private property inside and outside the buffer?

Many decisions about the types, configurations, and numbers of sites should be based on the needs of the species to be conserved. When such knowledge is lacking, it may be possible to gain insight by looking at related species or similar conservation contexts.

## Identify Threats to Candidate Buffers

As part of the planning process, it is essential to list, map, and quantify the current land uses and activities that predominate in the region. It is also essential to consider the economic and demographic changes occurring in the region and to anticipate how these changes will influence land use. Lists of compatible and incompatible uses should be made.

To determine whether a certain activity in a buffer area is compatible with conservation goals, the managing organization, in conjunction with concerned individuals and groups, should compile a checklist of factors that may contribute to ecological degradation and thus establish the level of threat in particular buffer areas. The geographic extent of impacts should be mapped, including detail on how these influences vary within the area (according to soils, slope, vegetation, riparian areas, trail or road density, and so on). Finally, it may be necessary to conduct research to assess what impacts will affect core areas or reduce the utility of the buffer zone—impacts such as an increase in nest parasites; changes in wind, temperature, relative humidity, and other abiotic variables at the edges of buffers or buffer/core boundaries; or interruption of wildlife movements. For example, Klein et al. (1995) conducted experiments to determine the relationship between the intensity of tourist disturbance and the negative effects on waterbirds in the Ding Darling National Wildlife Refuge of Florida. Experimental and observational studies could provide badly needed data for use in selecting sites and managing buffer areas.

Clearly, the evaluation of human uses within buffer areas is not a simple procedure. It is misleading, however, to view these different uses as equivalent with constant effects across all vegetation types and landforms. Consequently, these factors require a site-by-site evaluation. Ultimately, the evaluation of candidate buffer areas may require a draft management plan.

## Get Cooperation of Stakeholders

The management of ecological resources within and across buffer areas and buffer/core boundaries will only succeed if there is cooperation among landowners, elected officials, private organizations, and public land managers (Knight and Landres 1998). Agreement among stakeholders, therefore, is critical to the success of reserve networks (Tuxill and Nabhan 1998). Public and private participants seldom see eye to eye on ecological issues or sustainable uses of buffer areas, and there is little shared perspective on the implications of land uses in buffer zones for either ecological or economic goals.

Private landowners often resent interference in their use of their land and may be unwilling to cooperate in regional land-use decisions. They and their elected officials expect public lands to sustain local and regional economies largely through tourism and extractive uses of natural resources. Conservationists, on the other hand, often see development within buffer zones as anathema to the maintenance of biodiversity. Given these contrasting and often inflexible perspectives, there has been little basis for establishing common goals relating to buffer zones.

The establishment of conservation-compatible uses within buffer zones requires better cooperation among stakeholders. The solution entails setting up and maintaining a decision process that allows these stakeholders to clarify and secure their common interests by establishing cooperative mechanisms wherein appropriate information is available and used in planning, open discussion is used for consensus development, disputes are resolved constructively, and management policy is appraised on an ongoing basis (Clark 1992). Although the process can be costly in time and money, one product of such interactions is a map and plan in which all parties have a stake.

While development of shared goals relating to land issues is essential, it is only the first step. Beyond goals, effective implementation is needed. This requires sharing information across agencies and organizations and with diverse publics. Information should not be viewed as an end in itself but as a means to foster trust, cooperation, and effective conservation. Learning from experience is essential in this process (Clark 1996). Adaptive management—which includes organizational and policy learning as well as technical and individual learning—is one way to proceed. This kind of learning, however, does not happen on its own. It must be planned, managed, and rewarded.

In addressing issues that deal with buffer zones, all stakeholders must acknowledge their own responsibilities in land management. Individuals, businesses, and public officials whose interests are in some way connected to public lands must learn to work collaboratively with land-management agencies in developing effective decision-making and management processes.

## Incentives

Buffer and corridor planning, even when it is based on the best available ecological information, is a social exercise. What can biologists do to encourage "biodiversity-friendly" behavior among those who own, live in, exploit, and use areas that are adjacent to cores and linkages? How do we encourage dwellers and users of buffer zones to voluntarily adopt beneficial uses and avoid harmful ones in the buffer—without, that is, imposing legal mandates and offensive regulation?

An obvious answer is the use of positive incentives. With this in mind, it is important to consider a wide range of incentives for encouraging participation by various groups and individuals in buffer zone implementation. Diverse strategies for implementing buffer zones involve separate legislative, administrative, private, and restorative steps, often with different organizations or individuals taking the lead. A variety of incentives might be considered:

- *Aesthetic and recreational incentives.* People are encouraged to protect natural places that they are able to visit and enjoy for their natural beauty, solitude, or other aesthetic values. Recreational incentives include allowing and managing sustainable levels of hunting, hiking, fishing, camping and boating, or other pursuits compatible with buffer zone management plans.
- *Ethical and social incentives.* Some people will support buffer zone planning goals because they share an ethical belief that species and natural resources should be protected. Incentives for others may be the social pressures to cooperate with plans that promote public use areas and protect natural areas from development and exploitation.
- *Cultural and educational incentives.* Communities benefit from public or private land that is available for use as educational study sites—such as for field trips and living laboratories for learning. Communities may also realize the value of maintaining traditional ranching or other land-based lifestyles.
- *Economic incentives.* Economic incentives may be the most direct way of encouraging support from individuals or groups whose support is critical to buffer zone implementation. Examples of such incentives include:

1.  Conservation easements on private lands that result in tax benefits to individuals. Land trusts, conservation groups, and the wealthy obtain conservation easements on properties or purchase lands for protection within designated buffer zones to meet conservation goals and support the ecological integrity of the areas.
2.  Hunting fees to be collected by private landowners for access to their private lands for hunting.
3.  Ecotourism and recreation that generate income for the local and regional economy.
4.  Public relations incentives benefiting corporations and individuals who dedicate lands for protection and management suitable to the buffer zone.
5.  Small-scale harvesting of timber or other plant resources (herbs, mushrooms, medicinal plants) can lead to sustainable local economies in the absence of large-scale corporate harvesting operations.
6.  Encouragement of certain agricultural activities and cooperation in marketing between various groups involved in a buffer zone—predator-friendly and organic meat production, for example, or low-impact grazing as alternatives to traditional livestock farming.
7.  Reimbursement to ranchers for depredation of livestock by wildlife (as Defenders of Wildlife is doing for wolves).
8.  Highlighting the community benefits of sustainable harvests of fish, game, and plant products arising from adequate resource protection and enforcement.
9.  Publicizing the community benefits of soil conservation, watershed restoration, and protection from degraded water supplies.

It is imperative to showcase incentives that are successful—with the caveat that success means better protection of biodiversity, not merely economic gain. Demonstrating that sustainable development in the buffer zone can be economically viable yet ecologically compatible will encourage long-term investment in a region.

## Management of Buffer Zones

Every buffer needs a management plan. This plan should include an analysis of the prospects for regulation and management of use as well as the anticipated level of compliance from various groups. The planners should also question whether changes in the intensity or type of activities occurring in buffer zones enhance the buffer's utility for conservation goals. Moreover, they must consider the costs of monitoring, maintenance, mitigation, and enforcement.

To manage buffer zones within a regional reserve network, it is necessary to monitor activities and impacts. The information gathered during monitoring can be used in an adaptive management framework to establish practical measures of management success and guide decisions about land use in the future. The following checklist summarizes the necessary steps to sustain management of buffer areas within a reserve network:

1. *Document baseline attributes.* Select desired attributes of the buffer zone based on a compromise of the characteristics of the adjacent core zone and developed area.
2. *Monitor baseline attributes.* Measure attributes (abundance, distribution, size) with the best possible scientific methods.
3. *Analyze and determine changes.* Use GIS, sociopolitical surveys, and other analytical tools to analyze dynamics at various landscape scales.
4. *Select the species or process to monitor.* Select a keystone, indicator, or umbrella species, or select a process that is a critical indicator of ecosystem health for monitoring. Conduct a power analysis to see how much change could be detected given your budget (Stanley and Mills in press).
5. *Assess whether changes are positive or negative.* Determine trends and decide how much change is too much. Establish practical measures of management success. Beware of lag time.
6. *Determine management action.* Determine what actions will best maintain desired characteristics of the buffer ecosystem. Solicit input from NGOs, government agencies, and local residents.
7. *Enact management plans.* Carry out necessary management activities through an integrated effort by NGOs, government agencies, and local residents.

It is apparent from this list that the management plan for each buffer should incorporate a program for ongoing research to assess the compatibility of various practices in buffers. It is often a matter of degree, not an absolute, that makes certain uses of an area incompatible with conservation goals. Therefore, we must pay particular attention to the relationships between the gradients in intensity, duration, or frequency of activities and their effects on species and ecosystem processes.

As useful as it is to define compatible use, it is just as important to define the point at which a compatible activity becomes incompatible with the conservation goals of the buffer zone. In some cases, precise definitions may be possible—for example, most colony-breeding birds can tolerate tourist and other human traffic so long as people never approach nearer than 100 to 200 meters of the nesting colony (Rodgers and Smith 1995). In the majority of cases, however, it is unclear where the thresholds between compatibility and incompatibility lie—for example, it is difficult to define when a 10 percent increase in use will be enough to cross such a threshold. A further complication is that many activities have cumulative impacts that make the definition of separate thresholds even more difficult. Yet this is one of the most critical research needs regarding buffer zones. Clearly we need more information and examples of thresholds and nonlinearities.

## Reason for Optimism

Not much systematic thought has been devoted to the buffering of core conservation areas and networks. Certainly the subject is complex. But in addition, buffers have not been a traditional conservation element in North America. This pattern is likely to change, however, as habitat loss and fragmentation increase and as more information becomes available on the impacts of developed lands on protected areas.

Buffers, by their nature and by definition, are as much social as biological. Hence the prickly issue of human management is as relevant as the relatively straightforward problems of wildlife and land management. Planning and management of buffers are always more art than science. We have few guidelines for enhancing or even maintaining the conservation benefits of buffer areas. Yet the general guidelines reviewed here are flexible enough for use in a broad range of cases.

To develop cooperative stewardship in buffer zones we must keep in mind several points. First, we need to be clear about goals and work with all stakeholders to formulate an agreement on these goals. Second, we need to determine the range of activities that may occur in buffers. Third, we must characterize the potential impacts caused by an activity and determine the duration, frequency, and intensity of those impacts. Fourth, we must evaluate past and future trends before designing the reserve network. Finally, we must remember to offer incentives to push for greater popular support for protection of natural values and compatibility in resource use and to establish a dialogue about all these issues that will keep stakeholders committed to these projects.

The potential for buffer zones to enhance the conservation value of present and future reserves is still underappreciated. The opportunities are vast. And for most of the rare and threatened taxa in North America, an effective network that includes zones of cooperative stewardship may be their only hope. Much remains to be learned, of course, but if funders, institutions, and managers remain flexible and support good science and monitoring, we can be optimistic about the prospects.

# References

Andren, H., and P. Angelstam. 1988. Elevated predation rates as an edge effect in habitat islands: Experimental evidence. *Ecology* 69:544–547.

Angelstam, P. 1986. Predation on ground-nesting birds' nests in relation to predator densities and habitat edge. *Oikos* 47:365–373.

Baker, W. L, and R. L. Knight. 1999. Roads and fragmentation in the southern Rocky Mountains. In R. L. Knight, F. W. Smith, S. W. Buskirk, W. H. Romme, and W. L. Baker (eds.), *Forest fragmentation in the Southern Rocky Mountains*. Niwot: University Press of Colorado.

Bayne, E. M., and K. A. Hobson. 1997. Comparing the effects of landscape fragmentation by forestry and agriculture on predation of artificial nests. *Conservation Biology* 11:1418–1429.

Bierregaard, R. O. Jr., and P. C. Stouffer. 1997. Understory birds and dynamic habitat mosaics in Amazonian rainforests. In W. F. Laurance and R. O. Bierregaard Jr. (eds.), *Tropical forest remnants: Ecology, management, and conservation of fragmented communities*. Chicago: University of Chicago Press.

Bierregaard R. O. Jr., T. E. Lovejoy, V. Kapos, A. A. dos Santos, and R. W. Hutchings. 1992. The biological dynamics of tropical rainforest fragments. *BioScience* 42(11):859–866.

Blair, R. B. 1996. Land use and avian species diversity along an urban gradient. *Ecological Applications* 6:506–519.

Brown, J. H. 1988. Alternative conservation priorities and practices. Paper presented at the 73rd annual meeting of the Ecological Society of America, Davis, Calif.

Brown, M. T., J. M. Schaefer, K. H. Brandt, S. J. Doherty, C. D. Dove, J. P. Dudley, D. A. Eifler, L. D. Harris, R. F. Noss, and R. W. Wolfe. 1987. *An evaluation of the applicability of upland buffers for the wetlands of the Wekiva Basin*. Gainesville: Center for Wetlands, University of Florida.

Clark, T. W. 1992. Practicing natural resource management with a policy orientation. *Environmental Management* 16:423–433.

———. 1996. Learning as a strategy for improving endangered species conservation. *Endangered Species Update* 13(1–2):5–6 and 22–24.

Clark, T. W., and S. C. Minta. 1994. *Greater Yellowstone's future: Prospects for ecosystem science, management and policy*. Moose, Wyo.: Homestead Press.

Clarkson, D. A., and L. S. Mills. 1994. Ecological factors associated with hypogeous sporocarps in fragmented forests. *Northwest Science* 68:259–265.

Clergeau, P., J. P. L. Savard, G. Gennechez, and G. Falardeau. 1998. Bird abundance and diversity along an urban-rural gradient: A comparative study between two cities on different continents. *Condor* 100:413–425.

Cole, E. K., M. D. Pope, and R. G. Anthony. 1997. Effects of road management on movement and survival of Roosevelt elk. *Journal of Wildlife Management* 61:1115–1126.

Colwell, M. A., and S. L. Dodd. 1995. Waterbird communities and habitat relationships in coastal pastures of northern California. *Conservation Biology* 9:827–834.

Copeyon, C. K., J. R. Walters, and J. H. Carter III. 1991. Induction of red-cockaded woodpecker group formation by artificial cavity construction. *Journal of Wildlife Management* 55:546–556.

Cortner, H. J., P. D. Gardner, and J. G. Taylor. 1990. Fire hazards at the urban-wildland interface: What the public expects. *Environmental Management* 14: 57–62.

Crooks, K. 1997. Tabby go home: House cats and coyote interactions in southern California habitat remnants. *Wild Earth* 7(4):60–63.

Crumpacker, D. W., S. W. Hodge, D. Friedly, and W. P. Gregg. 1988. A preliminary assessment of the status of major terrestrial and wetland ecosystems on federal and Indian lands in the United States. *Conservation Biology* 2: 103–115.

Didham, R. K. 1997. The influence of edge effects and forest fragmentation on leaf litter invertebrates in central Amazonia. In W. F. Laurance and R. O. Bierregaard Jr (eds.), *Tropical forest remnants: Ecology, management, and conservation of fragmented communities*. Chicago: University of Chicago Press.

Dobkin, D. S., A. C. Rich, and W. H. Pyle. 1998. Habitat and avifaunal recovery from livestock grazing in a riparian meadow system of the northwestern Great Basin. *Conservation Biology* 12:209–221.

Duffus, D. A., and P. Dearden. 1990. Non-consumptive wildlife-oriented recreation: A conceptual framework. *Biological Conservation* 53:213–231.

Dustin, D. C., and L. H. McAvoy. 1987. Outdoor recreation and the environment: Problems and prospects. *Environmental Conservation* 9:343–346.

Florida Department of Agriculture and Consumer Services (FLDACS). 1993. *Silviculture best management practices*. Tallahassee: FLDACS.

Forbes, G. J., and J. B. Theberge. 1996. Cross-boundary management of Algonquin Park wolves. *Conservation Biology* 10:1091–1097.

Forman, R. T. T., D. S. Friedman, D. Fitzhenry, J. D. Martin, A. S. Chen, and L. E. Alexander. 1997. Ecological effects of roads: Toward three summary indices and an overview for North America. In K. Canters, A. Piepers, and D. Hendriks-Heersma (eds.), *Habitat fragmentation and infrastructure*. Delft: Ministry of Transport, Public Works, and Water Management.

Franklin, J. F. 1993. Preserving biodiversity: Species, ecosystems, or landscapes. *Ecological Applications* 3:202–205.

Friesen, L. E., P. F. J. Eagles, and R. J. Mackay. 1995. Effects of residential development on forest-dwelling neotropical migrant songbirds. *Conservation Biology* 9:1408–1414.

Gashwiler, J. S. 1970. Plant and mammal changes on a clearcut in west-central Oregon. *Ecology* 51:1018–1026.

Greenberg, C. H., D. G. Neary, and L. D. Harris. 1994. Effect of high-intensity wildfire and silvicultural treatment on reptile communities in sand-pine scrub. *Conservation Biology* 8:1047–1057.

Griggs, G. B., and B. L. Walsh. 1981. The impact, control, and mitigation of off-road vehicle activity in Hungry Valley, California. *Environmental Geology* 3:229–243.

Groom, M. J. 1992. Ecotourism in Manu Biosphere Reserve, Peru: Mitigating negative impacts on riverine species, In J. Kusler (ed.), *Ecotourism and resource conservation*. Berne, N.Y.: Association of Wetland Managers.

Harris, L. D. 1984. *The fragmented forest: Island biogeography theory and the preservation of biodiversity.* Chicago: University of Chicago Press.

———. 1989. The faunal significance of fragmentation of southeastern bottomland forests. In D. D. Hook and R. Lea (eds.), *Proceedings of the symposium: The forested wetlands of the southeastern United States.* General technical report SE-50. Ashville, N.C.: USDA Forest Service.

Harris, L. D., T. Hoctor, D. Maehr, and J. Sanderson. 1996. The role of networks and corridors in enhancing the value and protection of parks and equivalent natural areas. In R. G. Wright (ed.), *National parks and protected areas: Their role in environmental protection.* Oxford: Blackwell Science.

Heske, E. J. 1995. Mammalian abundances on forest-farm edges versus forest interiors in Southern Illinois: Is there an edge effect? *Journal of Mammalogy* 76(2):562–568.

Hummel, M. (ed.). 1989. *Endangered spaces: The future for Canada's wilderness.* Toronto: Key Porter Books.

Janzen, D. H. 1986. The eternal external threat. In M.E. Soulé (ed.), *Conservation biology: The science of scarcity and diversity.* Sunderland, Mass.: Sinauer.

Jensen, D. B., M. S. Torn, and J. Harte. 1993. *In our own hands: A strategy for conserving California's biological diversity.* Berkeley: University of California Press.

Johns, A. D. 1985. Selective logging and wildlife conservation in tropical rain forest: Problems and recommendations. *Biological Conservation* 31:355–375.

Jules, E. S. 1998. Habitat fragmentation and demographic change for a common plant: Trillism in old-growth forest. *Ecology* 79:1645–1656.

Jules, E. S., E. J. Frost, D. A. Tallmon, and L. S. Mills. In press. Ecological consequences of forest fragmentation: Case studies from the Klamath Mountains. *Natural Areas Journal.*

Kirkland, G. L. Jr. 1990. Patterns of initial small mammal community change after clearcutting of temperate North American forests. *Oikos* 59:313–320.

Klein, B. C. 1989. The effects of forest fragmentation on dung and carrion beetles (*Scarabaeinae*) communities in central Amazonia. *Ecology* 70:1715–1725.

Klein, M. L., S. R. Humphrey, and H. F. Percival. 1995. Effects of ecotourism on distribution of waterbirds in a wildlife refuge. *Conservation Biology* 9:1454–1465.

Knight, R. L. 1997. Field report from the new American West. In C. Meine (ed.), *Wallace Stegner and the continental vision.* Washington, D.C.: Island Press.

Knight, R. L., and T. W. Clark. 1998. Boundaries between public and private lands: Defining obstacles, finding solutions. In R. L. Knight and P. B. Landres (eds.), *Stewardship across boundaries*. Washington, D.C.: Island Press.

Knight, R. L., and P. B. Landres (eds.). 1998. *Stewardship across boundaries*. Washington, D.C.: Island Press.

Knight, R. L., G. N. Wallace, and W. E. Riebsame. 1995. Ranching the view: Subdivisions versus agriculture. *Conservation Biology* 9:459–461.

Laurance, W. F. 1994. Rainforest fragmentation and the structure of small mammal communities in tropical Queensland. *Biological Conservation* 69:23–32.

———. 1995. Extinction and survival of rainforest mammals in a fragmented tropical landscape. In William Z. Lidicker Jr. (ed.), *Landscape approaches in mammalian ecology and conservation*. Minneapolis: University of Minnesota Press.

Laurance, W. F., and C. Gascon. 1997. How to creatively fragment a landscape. *Conservation Biology* 11:577–579.

Lidicker, W. Z. 1995. *Landscape approaches in mammalian ecology and conservation*. Minneapolis: University of Minnesota Press.

Malcolm, J. R. 1997. Biomass and diversity of small mammals in Amazonian forest fragments. In W. F. Laurance and R. O. Bierregaard Jr. (eds.), *Tropical forest remnants: Ecology, management, and conservation of fragmented communities*. Chicago: University of Chicago Press.

Martell, A. M. 1983. Demography of southern red-backed voles *(Clethrionomys gapperi)* and deer mice *(Peromyscus maniculatus)* after logging in north-central Ontario. *Canadian Journal of Zoology* 61:958–969.

McCoy, M. B. 1994. Seasonal freshwater marshes in the tropics: A case in which cattle grazing is not detrimental. In G. K. Meffe and C. R. Carroll (eds.), *Principles of conservation biology*, 1st ed. Sunderland, Mass: Sinauer.

McIntyre, S., and S. Lavorel. 1994. Predicting richness of native, rare, and exotic plants in response to habitat and disturbance variables across a variegated landscape. *Conservation Biology* 8:521–531.

Mills, L. S. 1995. Edge effects and isolation: Red-backed voles on forest remnants. *Conservation Biology* 9:395–403.

———. 1997. Fragmentation of a natural area: Dynamics of isolation for small mammals on forest remnants. In R. G. Wright (ed.), *National parks and protected areas: Their role in environmental protection*. Cambridge, Mass.: Blackwell Press.

Noss, R. F., and A. Y. Cooperrider. 1994. *Saving nature's legacy: Protecting and restoring biodiversity*. Washington, D.C.: Island Press.

Petersen, M. R. 1996. Wilderness by state mandate: A survey of state-designated wilderness areas. *Natural Areas Journal* 16:192–197.

Raivio, S., and Y. Haila. 1990. Bird assemblages in silvicultural habitat mosaics in southern Finland during the breeding season. *Ornis Fennica* 67:73–83.

Ratti, J T., and K. P. Reese. 1988. Preliminary test of the ecological trap hypothesis. *Journal of Wildlife Management* 52(3):484–491.

Robinson, S. K., F. R. Thompson III, T. M. Donovan, D. Whitehead, and J.

Faaborg. 1995. Regional forest fragmentation and the nesting success of migratory birds. *Science* 267:1987–1990.

Rodgers, J. A. Jr., and H. T. Smith. 1995. Set-back distances to protect nesting bird colonies from human disturbance in Florida. *Conservation Biology* 9:89–99.

Rosenberg, D. K., K. A. Swindle, and R. G. Anthony. 1994. Habitat associations of California red-backed voles in young and old-growth forests in western Oregon. *Northwest Science* 68:266–272.

Schaefer, J. M., and M. T. Brown. 1992. Designing and protecting river corridors for wildlife. *Rivers* 3(1):14–26.

Shelford, V. E. 1933. Ecological Society of America: A nature sanctuary plan unanimously adopted by the society, December 28, 1932. *Ecology* 14:240–245.

Soulé, M. E., and M. Sanjayan. 1998. Conservation targets: Do they help? *Science* 279:2060–2061.

Soulé, M. E., D. T. Bolger, A. C. Alberts, R. Sauvajot, J. Wright, M. Sorice, and S. Hill. 1988. Reconstructed dynamics of rapid extinctions of chaparral-requiring birds in urban habitat islands. *Conservation Biology* 2:75–92.

Stanley, A. G., and L. S. Mills. In press. Statistical power of monitoring programs. *Wildlife Society Bulletin*.

Stanton, R. 1995. Managing liability exposures associated with prescribed fires. *Natural Areas Journal* 15:347–352.

Sullivan, T. P. 1979. Demography of populations of deer mice in coastal forest and clear-cut (logged) habitats. *Canadian Journal of Zoology* 57:1636–1648.

Thiollay, J.-M. 1995. The role of traditional agroforests in the conservation of rain forest bird diversity in Sumatra. *Conservation Biology* 9:335–353.

Torres, S. G., T. M. Mansfield, J. E. Foley, T. Lupo, and A. Brinkhaus. 1996. Mountain lion activity in California: Testing speculations. *Wildlife Society Bulletin* 24:451–460.

Tsuyuzaki, S. 1994. Environmental deterioration resulting from ski-resort construction in Japan. *Environmental Conservation* 21:121–125.

Tuxill, J., and G. P. Nabhan. 1998. *Plants and protected areas*. London: Stanley Thorns.

UNESCO. 1974. Task Force on Criteria and Guidelines for the Choice and Establishment of Biosphere Reserves. *Man and the biosphere report no. 22*. Bonn: UNESCO.

Warner, R. E. 1994. Agricultural land use and grassland habitat in Illinois: Future shock for midwestern birds? *Conservation Biology* 8:147-156.

# 8
## Why We Need Megareserves: Large-Scale Reserve Networks and How to Design Them

*John Terborgh and Michael E. Soulé*

Environmental excesses of the past and present have led to the current global extinction crisis. While much publicity has been given to the threat of habitat loss in the tropics, North America is by no means immune to threats of extinction. The number of species officially listed by the U.S. Department of Interior as "endangered" has already grown to over one thousand, and nearly five times this number are in line for listing. The specter of an endlessly expanding list has the U.S. Congress balking at renewing the Endangered Species Act.

Clearly we are approaching a turning point: a new policy for biodiversity will be enacted, but no one knows what that policy will be. If we stay on the same path—clear-cutting the last old-growth forests, overgrazing grasslands, fragmenting habitats, draining wetlands, polluting rivers—we know what will happen. Much of North America will lose every vestige of its wild origins, as has already happened over most of Europe. But to embark on a new path, we need a vision for the future. Do we want an America that is even more crowded, congested, and polluted than it is today? Do we want a continent that is wiped clean of old-growth forests and large carnivores, a continent that retains only remnants of its migratory birds, reptiles, amphibians, and native freshwater fish? Do we want to live in a continent of weeds?

Very few voters would answer these questions in the affirmative. North America north of Mexico has achieved a level of prosperity unimagined a century ago. For a large majority of citizens, material comfort is a fact of life. Yet the politicians who represent the North American people appear obsessed with a desire for more, rather than better. The quest for unending growth is the economic counterpart of the frontier mentality; today, however, consumer goods have replaced land, furs, and other raw natural resources. Isn't it time that North Americans started to enjoy their prosperity? Isn't it time to focus on quality rather than quantity? Everyone should be asking: "What kind of a continent do I want my children and grandchildren to live in?" Answering that question is what this book is about: quality of life, not just for a few, but for everyone; not just for human beings, but for all native species.

Is the future going to be simply an extension of the past—with ever more crowding, congestion, and resource depletion? Here we offer an alternative vision. It is a future in which humans are surrounded by beauty, spaciousness, and abundance. It is a future in which people derive spiritual nourishment from nature—a future in which we are no longer a destructive force but part of a larger whole. We look forward to a world in which the needs of future generations are respected as much as those of our own generation. We foresee a world in which there is freedom to live and enjoy, but not to destroy. Toward such a future, we sketch a roadmap. Many of the details are incomplete, but we see clearly where we want to go and we know roughly how to get there. We hope others will want to join us. The journey won't all be easy, but it is one well worth making.

Humans and nature can coexist, but peaceful coexistence cannot come about under present conditions. The revival and survival of nature across North America will require the establishment of a network of large nature reserves. Large areas managed for biodiversity are needed to ward off a host of ecological pathologies. Through conservation-oriented management of extensive core and multiple-use areas, the vital abiotic and biotic processes that sustain biodiversity can be perpetuated. Outside of biologically viable large reserves, ecological pathologies will continue to spread and take their toll.

Migratory songbirds, for example, have declined or disappeared from fragmented forests all over the continent. Nesting success is insufficient in many regions to compensate for adult mortality. The syndrome is so widespread that entire states (such as Illinois) have become population sinks for a host of common species (Robinson et al. 1996).

Alien species constitute another form of ecological pathology. Thousands of intentional and unintentional introductions have allowed aliens to become a pervasive presence in nearly every habitat and body of water on the continent. The situation in the Hawaiian Islands, where more than four thousand species of plants have been introduced, is even worse than on the mainland. When one notes that the number of native Hawaiian plants is only about 1100, the magnitude of the problem comes into focus. Stepped-up efforts to limit the entry of alien species have been partially effective, but new invaders appear every year nonetheless.

Alien species are a scourge. They dilute indigenous plant communities, alter the character of habitats, outcompete, kill, or eat native species, transmit diseases, and cause devastating blights. Alien species can be described as the ecological analog of cancer. And like cancer, many alien species have proved refractive to the best efforts of modern science in scores of unsuccessful efforts to contain, control, or eradicate them. The one generality that seems to apply is that aliens are slow to invade intact native communities on the land. But this observation offers little solace because intact native communities cover less than 10 percent of North America south of the boreal forest.

In the northern plains, poor nesting success of waterfowl is attributed to abnormally abundant nest predators and the invasion of Russian olive, an alien tree. Russian olive creates thick arboreal screens around prairie potholes, blocking the flight paths of diving ducks and eliminating otherwise prime sites as breeding habitat. In the southeast the bobwhite, once the region's most popular game bird, has become so rare that it is a treat nowadays to hear its clarion song. Superabundant "mesopredators" such as raccoons, opossums, and feral house cats are eliminating the bobwhite by imposing unsupportable levels of nest predation.

In large parts of the East, overabundant white-tailed deer are decimating acorn crops and tree seedlings—thereby altering tree recruitment patterns to an alarming degree (Alverson et al. 1994; McShea et al. 1997). Feral pigs in the South are equally destructive to forests and to the wildflowers that contribute 80 percent of the plant diversity of many temperate forests. Overbrowsing by ungulates, native and introduced, is so widespread that wildflowers are disappearing even in some of the most solicitously protected old-growth forests, such as the Heart's Content grove in Pennsylvania (Miller et al. 1992; Rooney and Dress 1997).

Huge areas of the Intermountain West have been invaded by cheatgrass, an exotic from southern Russia that destroys the rangeland for domestic livestock and wildlife alike. In the Southwest another aggressive alien, buffle grass, threatens to destroy the picturesque Sonoran Desert because it is flammable and can introduce fire to a system that is not adapted to it. Throughout the Midwest, inadvertently introduced zebra mussels are spreading rapidly in polluted streams, replacing native mollusks, and threatening some with extinction.

If all these trends and many others—such as pollution, overexploitation of useful species, and climate change—continue unchecked, the number of endangered species in the United States and elsewhere in North America will escalate until we are simply overwhelmed. The Endangered Species Act, for all its good intent, will become irrelevant to stanch the tide. We shall then discover with shock that we have lost the battle to conserve North America's biological heritage.

In the simplest terms, the battle against extinction is being lost because the processes that maintained biodiversity prior to human settlement have been disrupted. More than 90 percent of the land in the Lower 48 states has been logged, plowed, mined, overgrazed, paved, or otherwise modified from the presettlement condition. Fire suppression has altered the composition of plant communities in nearly every state (Leach and Givnish 1996). Top predators have been extirpated or reduced to scattered populations throughout much of the continent. Rivers have suffered even worse insults, having been dammed, channeled, diked, and converted to open conduits for human, agricultural, and industrial wastes (Dynesius and Nilsson 1994). Accordingly, aquatic organisms are under siege to a much greater extent than terrestrial life (Abramovitz 1996).

If North America is not to lose species by the hundreds or even thousands in the twenty-first century, it will be imperative to heed Aldo Leopold's aphorism that the intelligent tinkerer saves all the pieces. That policy would be hard enough in the best of cases, but we have made matters worse by losing some pieces already (extinct species) and by having to cope with thousands of others that do not fit (alien species). Putting nature back together will not be easy. The least of the challenges is that scientists do not know how to do it, at least not in detail. But scientists do know some of the things that must be done.

Here are some of the crucial requirements. More land must be protected (Chapter 5). Land use must be prescribed on very large spatial scales (Chapter 2). Management practices must be reformed (Chapters 6 and 7). Top carnivores must be restored in many places where they have been extirpated (Chapter 3). Alien species must be combatted as a matter of public policy (Chapter 4). Disturbed and degraded habitats must be made more natural (Chapter 4). Free run must be given to physical processes, such as wildfires and floods, that rejuvenate plant communities and shape the landscape (Chapter 2).

The vision propounded throughout this book is the goal of bringing back wildness in North America by healing the wounds of past excesses and indifference. A more specific objective is to ensure the permanence of all native species by providing ecological conditions that will sustain them indefinitely. Some will dismiss this vision as impossible, a flight of fancy, too idealistic. But the framers of the U.S. Constitution were remarkably idealistic, and yet no one criticizes their idealism now. What we propose is neither impossible nor impractical. The restoration of wild nature in North America is not a flight of fancy. If the people so desire, and act according to that desire, it can happen.

The environment is being mistreated and scientists know much that most to be done to fix it. Here we describe what it will take. On the social side, the key ingredient is people who are not afraid to dream of a better world. The rest is science—and that is the easy part, even if there are questions still to be answered. At the heart of the science is the restoration of the abiotic and biotic processes that sustained biodiversity over the millennia prior to the advent of humans. The leading abiotic processes essential to maintaining biodiversity are unrestrained fire and flood regimes. Modern forestry practices include widespread fire suppression and the controlled use of fire to limit hardwood regeneration in pine stands. Since fire-dependent plant communities disappear under these conditions, fire regimes that mimic the frequencies and intensities prevalent in presettlement times will need to be reinstituted. Similarly, prairies resembling those found by the pioneers can be recreated only by restocking native herbivore assemblages and allowing wildfires to open thick grass swards to colonization of native forbs and other species. Restoring natural flows to rivers will entail the removal of dams, dikes, and other water control structures, accompanied by efforts to reduce silt loads and pollution. The Kissimmee River in Florida is already undergoing an extensive (and expensive) restoration in recognition of past mistakes. Restoration of other rivers is under discussion or in the planning stage.

Among the essential biotic processes that regulate biodiversity are species interactions such as predation, pollination, parasitism, seed dispersal, seed predation, and herbivory. When these processes veer out of their "normal" ranges in response to fragmentation, habitat degradation, exotic species, and the absence of key members of the native fauna, a cascade of biological effects is unleashed—leading to what are termed "secondary extinctions." The challenge of restoring all these processes to their presettlement states presents a serious scientific problem. Nevertheless, there is abundant evidence (reviewed in Chapter 3) to support the notion that top-down regulation is essential to stabilizing biotic interactions. Top-down forces are those exerted by species at one trophic level on those at the next lower level: effects of predators on herbivores, for example, and herbivores on plants. Where some or all of the large carnivores have been extirpated (in more than 90 percent of the Lower 48 states), top-down forces have been drastically weakened. The consequence has been widespread eruptions of herbivores—such as beavers and white-tailed deer—and "mesopredators" such as foxes, raccoons, and opossums, species whose numbers were once maintained at lower levels by larger predators. In the absence of large predators, some of these smaller animals have become potent agents of secondary extinction. Reintroduction of native predators, therefore, must stand at the top of the agenda in any effort to promote the recovery of partially degraded ecosystems.

Restoration of natural abiotic and biotic process regimes will also take lots of land—perhaps 30 to 40 percent of the national territories of the United States. and Canada. Such a figure may sound preposterous. But in fact it is realistic both scientifically and socially. This is because much of the needed land is already in the public sector. Consider the United States. Roughly 40 percent of the national territory is public land, most of it federal, excluding reservation lands of Native Americans. If federal and state land management policies could be made more compatible with biodiversity conservation, much of the national estate could eventually be incorporated into either core protected areas or buffers. It is a matter of persuading politicians to expand current roadless and wilderness areas and then restoring the connections between them.

We hasten to stress that creating networks of reserves out of large tracts of public land does not imply "locking up" natural resources. The resources of reserves can still be used (except in core areas). What must change are management practices. Today many public lands are being abused for the benefit of the few at the expense of the many. Is the public going to stand passively by while federal, provincial, and state agencies continue to authorize mining operations near national parks, overgrazing of public lands, clear-cutting of steep slopes, below-cost logging on public lands, destruction of trout and salmon streams, poisoning of prairie dogs, draining of wetlands, and a myriad of other environmental atrocities? Or are the citizens of North America finally going to wake up and see what is being done to the natural beauty of the land and its precious resources? The current policies are anachronisms—vestiges of an outmoded frontier mentality. They favor short-term gain over long-term prosperity and by doing so are despoiling America the Beautiful.

Reforming land management policies of federal, state, and provincial agencies is only one route to attaining the vision of a restored wild America. There are, in addition, abundant opportunities for private initiatives. Wildlands are available to be bought, for example, in northern Maine. And there is money to buy them. The current economic boom has generated trillions of dollars of wealth concentrated in the hands of a few. Over the last several decades individuals and foundations have collectively purchased millions of acres in the United States and elsewhere. Doug Tompkins, to cite one example, is offering to help create a major new national park in Chile and to preserve the last alerce trees, the redwoods of the Southern Hemisphere. Ted Turner is buying land for the protection of nature in Argentina and the United States, as well. Private philanthropy holds the possibility for establishing core areas and key corridors in ecosystems not well represented in public landholdings.

There is nothing new in principle about the idea of reserve networks. The idea has been around for a long time. As we have seen, the components already exist in the United States, Canada, and elsewhere. Our definition of a reserve network is a land management unit large enough to contain viable populations (at least several hundred individuals) of all native species (Chapters 1 and 5). In the Rocky Mountains, grizzly bears and wolverines would set the minimum; in other parts of the continent it might be wolves, pumas, or jaguars. The absolute size of reserve networks will vary. What is important is that native species have the space and conditions they need to survive over the long run.

Reserve networks will be designed around core areas afforded the highest level of protection, where most mechanized and commercial activities are excluded (Chapter 5). The last and best remaining examples of unspoiled North America should be preserved in cores. Recovered areas will have to serve where large tracts of pristine habitat no longer exist. Many core areas can be designated around existing wilderness in national parks, national forests, Crown lands, BLM lands, military reservations, state parks, Nature Conservancy holdings, or private reserves.

Cores will be protected by "buffer zones"—areas to which less stringent conservation criteria apply (Chapter 7). Many public lands managed for multiple use can serve as buffers. The purpose of buffers is to shield cores from pernicious external influences, such as alien species, and to expand the area of habitat available to species tolerant of some subsistence or commercial activities, such as light grazing or selective logging. Full implementation of reserves will require reforms in timber and grazing management policies, but such reforms will have to emerge from the political arena and may take time. This is not to say that current policies, however defective, should be an excuse for delay in planning and implementing reserve networks.

Corridors form the third and last architectural component of reserve networks, linking cores and buffers to one another (Chapter 6). Corridors may be many things. They might be private ranches with conservation easements, say, to fill the gaps between national forests and other public lands, so that grizzlies and wolves may enjoy safe passage between secure redoubts. In the East, even a mosaic of fields and woodlots might serve for the safe passage of certain wide-ranging species. The point of corridors is to provide thoroughfares for the mobile elements of nature so that separate cores and buffers do not become demographic and genetic islands.

Corridors are of special importance for the larger members of our fauna. Large animals tend to be rare and often move greater distances than smaller fauna. A single wolf pack may use hundreds of square kilometers, for example, perhaps an entire core. But a wolf pack has only one breeding male and female, hence genetically it is but a single pair. So that young wolves may reproduce without inbreeding, they must be able to disperse to the territory of the next pack to find potential mates. To do this, they need secure passageways through which they can travel without undue risk of meeting trigger-happy hunters, irate sheepherders, or death under an 18-wheeler on the interstate.

Corridors are a new concept in conservation biology, and in many respects they are untested. We know, for example, that young wolves have to move to find mates, often over large distances. But we don't know enough about how they move, what kinds of places they prefer, and what kinds they avoid. Several species of large mammals are being studied with radio collars to find out more about their wanderings and what habitats they use at different seasons. Elk have seasonal migrations between their high-country summer quarters and valley-floor winter ranges. Many, but not all, elk migrations take place entirely within federal lands in the West. Where elk have to cross private lands, fences and highways can be barriers or sources of mortality. Landowner incentives already exist to reduce such conflicts, and more will be needed.

Smaller creatures may need or require corridors, too, but on smaller scales. Aquatic turtles are at risk when they have to cross roads in search of nesting sites. Some species of snakes concentrate at dens for the winter and then fan out over thousands of hectares during the summer before returning to the den in the fall. Roads are unkind to snakes, too. Many butterflies fly in reference to landmarks, following hedgerows, streams, or forest edges from one patch of suitable habitat to another. Many frogs and some salamanders migrate seasonally to breeding ponds, dispersing widely at other times. The needs of all these animals, and many others, could be served by appropriately restoring the landscape.

When we say "appropriately restoring the landscape," we are not imagining an elaborate set of regulations and a vast bureaucracy to administer them. We are imagining a more eco-friendly society, one that has made an ethical decision to protect nature. Farmers could express their sympathy for wildlife by reestablishing the hedgerows that were so prevalent fifty years ago. Many ranchers already contribute by restricting access of their cattle to riparian corridors and by taking a more relaxed attitude toward large carnivores. Timber companies could do their part by adopting, where practical, uneven-age management, by obliterating logging roads and by sparing den trees and snags. And highway departments could reduce the carnage of roadkills by fencing rights-of-way, building animal underpasses, and avoiding the use of solid concrete barriers to separate oncoming lanes of traffic. Few of these measures would detract significantly from the income of farmers, ranchers, or loggers, or add noticeably to the burden of taxpayers.

Policies to protect watersheds and stream corridors benefit both people and living nature. Preliminary steps toward such policies have been taken in parts of the United States for the purpose of improving water quality. In the Chesapeake Bay watershed, for example, building permits are not issued for construction in floodplains of permanent streams. These and other associated regulations, codified in a tristate agreement, have resulted in notably cleaner tributaries and a cleaner, more productive, bay. But in other parts of the country, clearcuts, cattle pastures, cultivated fields, housing developments, and parking lots extend to the very rims of streambanks, so that contaminated runoff drains directly into aquatic habitats and into the water supply. Does anyone really want to drink water that has drained off a cattle pasture, golf course, or industrial complex? Improved management of stream corridors will come about through public pressure and civic action, quite independently of any value stream corridors may have for wildlife.

Proper protection of watersheds, stream corridors, and associated wetlands could yield major ancillary benefits to wildlife. Floodplains and wetlands are the most productive elements of the landscape, supporting higher densities of most groups of animals than drought-prone uplands. Stream valleys are natural biological connectors used by aquatic and terrestrial forms alike to move from one part of a landscape to another. The inherently dendritic form of drainage patterns ensures interconnectedness throughout a watershed as well as wide mobility to creatures traveling in or along streams. More enlightened policies toward streams and watersheds could go a long way toward creating movement corridors for animals and the needed interconnectedness on a landscape-watershed level. Connectivity between watersheds to ensure passage of migratory, nomadic, and wide-ranging species is an important feature of reserve design. Additional needs for interconnectedness could be met on a case-by-case basis as the large-scale movements of animals are better understood.

## Coda

More than a century ago, our forebears completed the conquest of the frontier. For a century and more before that, they waged war against wild America so successfully that they almost extinguished it. Mop-up operations against nature have characterized the twentieth century. Now we are seeing the folly in these past excesses. The loss of wild America diminishes our quality of life. How many of us would be pleased to live in a land without songbirds? In a land without wildflowers? In a land without majestic forests and windswept prairies? Wild nature is worth having because it enriches our lives and nourishes our psyches.

This book has described how the restoration of wild America can be accomplished through the establishment of a continental system of reserve networks constructed of cores, buffers, and corridors. Reserve networks can be designed in the Far North from existing wildlands without the need for major biological restoration. Elsewhere—that is, over most of the continent—wildlands will have to be recreated through a program of adaptive management. The goal is to restore, over large portions of the continent, the abiotic and biotic processes that sustain biodiversity. Essential processes include fire and flooding that shape the physical environment, predation, movements such as migration and dispersal, and others that define the interactions between plants and animals. This restoration implies not merely the qualitative reestablishment of such processes but the quantitative reinstatement of the homeostatic mechanisms that stabilize natural biotic communities and help them resist invasion by exotics. The recovery of the North American continent (except for parts of northern Canada and Alaska) thus presents major challenges, but challenges that fall largely within current scientific capability. Beyond science, what we need most is the political will to succeed in an exciting venture that will ensure a better future for all.

## References

Abramovitz, J. N. 1996. Imperiled waters, impoverished future: The decline of freshwater ecosystems. *Worldwatch Paper 128*. Washington, D.C.: The Worldwatch Institute.

Alverson, W. S., W. Kuhlmann, and D. M. Waller. 1994. *Wild forests: Conservation biology and public policy*. Washington, D.C.: Island Press.

Dynesius, M., and C. Nillson. 1994. Fragmentation and flow regulation of river systems in the northern third of the world. *Science* 266:753–762.

Leach, M. K., and T. J. Givnish. 1996. Ecological determinants of species loss in remnant prairies. *Science* 273(5281):1555–1558.

McShea, W. J., H. B. Underwood, and J. H. Rappole. 1997. *The science of overabundance: Deer ecology and population management*. Washington, D.C.: Smithsonian Institution Press.

Miller, S. G., S. P. Bratton, and J. Hadidian. 1992. Impacts of white-tailed deer on endangered and threatened vascular plants. *Natural Areas Journal* 12:67–74.

Robinson, S. K., F. R. Thompson III, T. M. Donovan, D. R. Whitehead, and J. Faaborg. 1996. Regional forest fragmentation and the nesting success of migratory birds. *Science* 267:1987–1990.

Rooney, T. P., and W. J. Dress. 1997. Species loss over sixty-six years in the ground-layer vegetation of Heart's Content, an old-growth forest in Pennsylvania, USA. *Natural Areas Journal* 17:297–305.

# About the Contributors

HÉCTOR ARITA is research professor at the Institute of Ecology of the National University of Mexico (UNAM). His research interests include the conservation biology of tropical mammals and the large-scale (continental) patterns of the distribution of biodiversity. He is coeditor of the Mexican National Atlas of Mammals, to be published by the Mexican Commission on Biodiversity.

DIANE BOYD-HEGER has studied the ecology of wolves and other large carnivores since 1977 in Minnesota, Montana, British Columbia, Alberta, the Northwest Territories, and Romania. She presently works for the Arizona Department of Game and Fish as the field supervisor for the Mexican wolf reintroduction program.

ERIC DINERSTEIN has studied the biology of large mammals (rhinoceros and tigers) in Asia and the ecology of fruit-eating bats in Latin American tropical forests. He is now chief scientist at World Wildlife Fund and is mainly concerned with setting conservation priorities within the most biologically important ecoregions of the world.

RODOLFO DIRZO is professor and researcher at the Institute of Ecology, National University of Mexico (UNAM). His main interests include evolutionary ecology, tropical ecology, and biological conservation, particularly the consequences of habitat fragmentation and defaunation in the tropics. He teaches field ecology and biological conservation in Mexico, Costa Rica, and Brazil.

DAN DOAK is interested in several topics of basic and applied ecology, including metapopulations, plant-herbivore interactions, and stability–diversity relationships. He has developed and applied demographic techniques to address conservation issues involving both rare plants and vertebrates including grizzlies, spotted owls, desert tortoises, sea otters, and cheetahs.

ANDY DOBSON teaches ecology, epidemiology, and conservation biology at Princeton University. His research interests include parasites in game birds in Scotland, elephants in Africa, brucellosis in Yellowstone, and the role that infectious disease plays in natural ecological systems.

JAMES A. ESTES is interested in community- and ecosystem-level effects of consumer-prey interactions. His research has focused on the sea otter/ kelp forest system, although he also has studied a variety of other marine species and systems. He is a research biologist with the U.S. Geological Survey and an adjunct professor at the University of California at Santa Cruz.

MERCEDES FOSTER is a research zoologist with the Patuxent Wildlife Research Center (USGS) at the National Museum of Natural History in Washington. She studies ecology and behavior of neotropical birds and plants and has been active in conservation efforts in Latin America. She heads the Biological Diversity Handbook Project to produce handbooks of standard methods for measuring and monitoring biodiversity.

STEVE GATEWOOD is a naturalist with a diverse natural resources background working for state and local agencies, academic institutions, non-profit organizations, and as a private consultant. His interests include plant community ecology and mapping, watershed planning, wetland restoration, preserve management, reserve design, and project management. He is currently executive director of The Wildlands Project.

BARRIE GILBERT is a senior scientist in the Conservation Biology Program at Utah State University. His interests include the behavior and ecology of large mammals and the role of dense populations of bears in transporting nutrients to watersheds above salmon-spawning streams. Since 1988, Dr. Gilbert has served on the science council of the Greater Yellowstone Coalition.

MICHAEL GILPIN, who came to conservation biology via the theoretical sides of physics, operations research, and community ecology, has largely been interested in questions concerning single species and their viability, recovery, and restoration. He is professor of biology at the University of California at San Diego.

MARTHA GROOM recently moved from North Carolina State University to the University of Washington. Her work has focused on the influence of habitat fragmentation and prior land-use history on population and community dynamics, including studies of the effects of spatial dispersion on extinction risk of plant populations, and avian ecology and conservation in Latin America.

DEBORAH B. JENSEN is a vice president and the director of the conservation science division of The Nature Conservancy. She is responsible for all the scientific activities of TNC's domestic and international programs. Her first book, *In Our Own Hands: A Strategy for Conserving Biodiversity in California* (Jensen, Torn, and Harte), was published by the University of California Press in 1993. The focus of her research has been forest ecology.

DAVID JOHNS, a political scientist, is a cofounder and was first president of The Wildlands Project. He continues to serve on the boards of TWP and the journal *Wild Earth*. His conservation work is now focused on regional wildlands reserve design in the Klamath-Siskiyou and Yellowstone-to-Yukon regions. He studies the integration of science, advocacy, and eco-centric values, and works on strategies for implementation of reserve designs.

RICHARD L. KNIGHT teaches wildlife conservation at Colorado State University. Recent books he has edited include *Stewardship Across Boundaries* and *A New Century for Natural Resources Conservation*, both published by Island Press. He and his wife, Heather, practice community-based conservation in northern Colorado. His research interests focus on the ecological effects of outdoor recreation on public lands, and the conversion of farm and ranch lands to rural housing developments..

CARLOS MARTÍNEZ DEL RIO teaches physiological ecology, ecology, and ornithology at the University of Arizona. His research interests include animal–plant interactions and the role that physiological traits play in shaping them. He teaches field ecology courses in Chile, Colombia, and Mexico. He currently works and plays between the deserts and the mountains of the Sonoran region.

DAVID MATTSON is a wildlife biologist with the U.S. Geological Survey at the University of Idaho. He has studied grizzly bears and the ecosystems they live in for the last twenty years. His research has included the development of models of grizzly bear habitat relations at all scales.

BRIAN J. MILLER is a conservation biologist for the Denver Zoo. He has studied black-footed ferrets, and has recently been studying jaguars and pumas in western Mexico. He is on the board of directors of The Wildlands Project and is a Smithsonian Research Associate.

L. SCOTT MILLS teaches wildlife biology at the University of Montana. His research interests include the causes and consequences of forest fragmentation and the viability of vulnerable populations of animals, with an emphasis on the interface between genetics and ecology. His current field projects focus on red-backed voles, mycorrhizal fungi, snowshoe hares, and Canada lynx.

LISA MILLS is the director of the Montana Natural History Center, a nonprofit conservation education organization based in Missoula, Montana. She is also a board member of the Crown of the Continent Ecosystem Education Consortium, which actively promotes public participation in biodiversity issues in the northern U.S. Rockies and Canada.

ELLIOTT A. NORSE is a marine and forest conservation biologist. He is president of Marine Conservation Biology Institute (MCBI), an organization dedicated to advancing the new science of marine conservation biology. He has worked for the U.S. Environmental Protection Agency, the President's Council on Environmental Quality, the Ecological Society of America, and The Wilderness Society.

REED F. NOSS is coexecutive director of the Conservation Biology Institute, an international consultant in conservation, and science editor for *Wild Earth*. From 1993 through 1997 he was editor of *Conservation Biology* and is currently president-elect of the Society for Conservation Biology.

PAUL PAQUET is an adjunct professor at the University of Calgary, Alberta, and senior wildlife ecologist for the Conservation Biology Institute. He also coordinates World Wildlife Fund's Large Carnivore Strategy for the Rocky Mountains. Since 1972, he has studied the ecology of wolves, coyotes, and bears in Arizona, Oregon, Ontario, Manitoba, Alberta, British Columbia, Northwest Territories, Slovakia, Poland, and Italy. Recent research has focused on the behavior of large carnivores in human-dominated landscapes.

KATHERINE RALLS is a research zoologist with the National Zoological Park, Smithsonian Institution. She has worked on inbreeding depression in zoo populations and several threatened and endangered species including San Joaquin kit foxes, California sea otters, and Hawaiian monk seals, and has served on numerous scientific advisory committees for government agencies.

SADIE RYAN is a former research assistant to A. P. Dobson. She received her bachelor's degree from Princeton University, where her thesis research focused on the interaction between primate social systems and the transmission of infectious diseases. Her interests and past research include primatology, disease ecology, conservation, and neotropical ecology. Currently she is a researcher for the Conservation Program at the Lincoln Park Zoo in Chicago.

J. MICHAEL SCOTT is a research biologist with the U.S. Geological Survey and professor of wildlife at the University of Idaho. He has authored papers on conservation and ecology of endangered Hawaiian forest birds, recovery of endangered species, reserve identification, selection, and design, and the use of Landsat data for predicting species occurrences.

DANIEL SIMBERLOFF is a professor in the department of ecology and evolutionary biology at the University of Tennessee. His interests include invasion biology, community ecology, island biology, and biogeography. He recently coedited *Strangers in Paradise,* a book on the impact and management of introduced species in Florida, published by Island Press.

MICHAEL E. SOULÉ is a past president and current science director of The Wildlands Project. His interests in conservation include genetics, the regulation of ecosystems, and the science/policy interface. He was a founder of the Society for Conservation Biology and The Wildlands Project. He has taught at the Universities of California at San Diego and Santa Cruz, where he is Professor Emeritus, and at the Universities of Michigan and Malawi.

JOHN TERBORGH is James B. Duke Professor of Environmental Science at Duke University and codirector of the Duke University Center for Tropical Conservation. His research falls in the area of community ecology and currently emphasizes plant–animal interactions and the role of "top-down" processes in structuring ecosystems. Long experience in the Amazon has prompted an interest in the special challenges of conserving tropical biodiversity.

STEVE TROMBULAK teaches conservation biology and vertebrate ecology at Middlebury College in Vermont. His research interests include the winter ecology of bats, the design of ecological reserve systems, and the use of conservation biology in the development of policies for the management of forest ecosystems.

# Index